Palgrave Advanced Texts in Econometrics

Series Editor
Michael Clements
ICMA Centre, Henley Business School
University of Reading
Wheatley, UK

Palgrave Advanced Texts in Econometrics is a series that provides coverage of econometric techniques, applications and perspectives at an advanced research level. It will include research monographs that bring current research to a wide audience; perspectives on econometric themes that develop a long term view of key methodological advances; textbook style presentations of advanced teaching and research topics. An over-riding theme of this series is clear presentation and accessibility through excellence in exposition, so that it will appeal not only to econometricians, but also to professional economists and, particularly, to Ph.D students and MSc students undertaking dissertations. The texts will include developments in theoretical and applied econometrics across a wide range of topics and areas including time series analysis, panel data methods, spatial econometrics and financial econometrics.

More information about this series at
http://www.palgrave.com/gp/series/14722

Hossein Hassani · Rahim Mahmoudvand

Singular Spectrum Analysis

Using R

Hossein Hassani
Research Institute of Energy
  Management and Planning
University of Tehran
Tehran, Iran

Rahim Mahmoudvand
Department of Statistics
Bu-Ali Sina University
Hamedan, Iran

Palgrave Advanced Texts in Econometrics
ISBN 978-1-137-40950-8   ISBN 978-1-137-40951-5 (eBook)
https://doi.org/10.1057/978-1-137-40951-5

Library of Congress Control Number: 2018941884

© The Editor(s) (if applicable) and The Author(s) 2018
The author(s) has/have asserted their right(s) to be identified as the author(s) of this work in accordance with the Copyright, Designs and Patents Act 1988.
This work is subject to copyright. All rights are solely and exclusively licensed by the Publisher, whether the whole or part of the material is concerned, specifically the rights of translation, reprinting, reuse of illustrations, recitation, broadcasting, reproduction on microfilms or in any other physical way, and transmission or information storage and retrieval, electronic adaptation, computer software, or by similar or dissimilar methodology now known or hereafter developed.
The use of general descriptive names, registered names, trademarks, service marks, etc. in this publication does not imply, even in the absence of a specific statement, that such names are exempt from the relevant protective laws and regulations and therefore free for general use.
The publisher, the authors, and the editors are safe to assume that the advice and information in this book are believed to be true and accurate at the date of publication. Neither the publisher nor the authors or the editors give a warranty, express or implied, with respect to the material contained herein or for any errors or omissions that may have been made. The publisher remains neutral with regard to jurisdictional claims in published maps and institutional affiliations.

Cover illustration: Pattern adapted from an Indian cotton print produced in the 19th century

Printed on acid-free paper

This Palgrave Pivot imprint is published by the registered company Macmillan Publishers Ltd. part of Springer Nature
The registered company address is: The Campus, 4 Crinan Street, London, N1 9XW, United Kingdom

Preface

Time series analysis is crucial in the modern world as time series data emerge naturally in the field of statistics. As a result, the application of time series analysis covers diverse areas, including those relating to ecological and environmental data, medicine and more importantly economic and financial time series analysis. In the past, time series analysis was restricted by the necessity to meet certain assumptions, for example, normality. In addition, the presence of outlier events, such as the 2008 recession, which causes structural changes in time series data, has further implications by making the time series non-stationary. Whilst methods have been developed using condemning time series models, such as variations of autoregressive moving average models, ARIMA models, such methods are largely parametric. In contrast, Singular Spectrum Analysis (SSA) is a non-parametric technique and requires no prior statistical assumptions such as stationarity or linearity of the series and works with both linear and non linear data. In addition, SSA has outperformed methods such as ARIMA, ARAR and Holt-Winters in terms of forecast accuracy in a number of applications. The SSA method consists of two complementary stages, known as decomposition and reconstruction, and both stages include two separate steps. At the first stage the time series is decomposed and at the second stage the original series is reconstructed and this series, which is noise free, is then used to forecast new data points. The practical benefits of SSA have resulted in its wide using over the last decade. As a result, the successful applications of SSA can now be identified across varying disciplines such as physics, meteorology, oceanology, astronomy, medicine, climate data, image processing, physical sciences, economics and finance. Practically there are few programs, such as SAS and Caterpillar, which allow performing the SSA technique, but these require payments which are sometimes not economical for an individual researcher. R is an open-source software package that was developed by Robert Gentleman and Ross Ihaka at the University of Auckland in 1999. Since then, it has experienced a huge growth in popularity within a short span of time. R is a programme which allows the user to create their own objects, functions and packages. The R system is command driven and it documents the analysis steps making it easy to reproduce or update the analysis and figure errors. R can be installed on any platform and is license free. A major advantage with R is that it allows integrating and interacting with other paid platforms such as SAS, Stata, SPSS and Minitab. Although there are some books in the market relating to SSA, this book is unique as it not only details the theoretical aspects underlying SSA, but also provides a comprehensive guide enabling the user to apply the theory in practice using the R software. This book provides the user with step-by-step coding and guidance for the practical application of the SSA technique to analyse their time series databases using R. We provided some basic R commands in Appendix, so the readers who are not familiar with this language please learn the very basics in the Appendix at first.

The help of Prof. Kerry Patterson and Prof. Michael Clements in editing the text is gratefully acknowledged. Discussions with Kerry and Michael helped to clarify various questions treated on the following pages. We thank both for their encouragement.

As this book endeavours to provide a concise introduction to SSA, as well as to its application procedures to time series analysis, it is mainly aimed at masters and Ph.D.'s students with a reasonably strong stats/maths background who wants to learn SSA, and is already acquainted with R. It is also appropriate for practitioners wishing to revive their knowledge of times series analysis or to quickly learn about the main mechanisms of SSA. On the time series side, it is not necessary to be an expert on what is popularly called Box-Jenkins modelling. In fact this could be a disadvantage since SSA modelling start from a somewhat different point and in doing so challenges some of the underlying assumptions of the Box-Jenkins approach.

Tehran, Iran Hossein Hassani
Hamedan, Iran Rahim Mahmoudvand
June 2018

Contents

1 **Univariate Singular Spectrum Analysis**   1
    1.1 Introduction   1
    1.2 Filtering and Smoothing   3
    1.3 Comparing SSA and PCA   13
    1.4 Choosing Parameters in SSA   14
       1.4.1 Window Length   15
       1.4.2 Grouping   22
    1.5 Forecasting by SSA   27
       1.5.1 Recurrent Forecasting Method   29
       1.5.2 Vector Forecasting Method   30
       1.5.3 A Theoretical Comparison of RSSA and VSSA   31
    1.6 Automated SSA   33
       1.6.1 Sensitivity Analysis   36
    1.7 Prediction Interval for SSA   38
    1.8 Two Real Data Analysis by SSA   40
       1.8.1 UK Gas Consumption   40
       1.8.2 The Real Yield on UK Government Security   44
    1.9 Conclusion   47

2 **Multivariate Singular Spectrum Analysis**   49
    2.1 Introduction   49
    2.2 Filtering by MSSA   50

		2.2.1 MSSA: Horizontal Form (HMSSA)	50
		2.2.2 MSSA: Vertical Form (VMSSA)	59
	2.3	Choosing Parameters in MSSA	64
		2.3.1 Window Length(s)	65
		2.3.2 Grouping Parameter, r	66
	2.4	Forecasting by MSSA	68
		2.4.1 HMSSA Recurrent Forecasting Algorithm (HMSSA-R)	68
		2.4.2 VMSSA Recurrent Forecasting Algorithm (VMSSA-R)	71
		2.4.3 HMSSA Vector Forecasting Algorithm (HMSSA-V)	75
		2.4.4 VMSSA Vector Forecasting Algorithm (VMSSA-V)	77
	2.5	Automated MSSA	79
		2.5.1 MSSA Optimal Forecasting Algorithm	79
		2.5.2 Automated MSSA R Code	80
	2.6	A Real Data Analysis with MSSA	82

3 Applications of Singular Spectrum Analysis — 87
- 3.1 Introduction — 87
- 3.2 Change Point Detection — 88
  - 3.2.1 A Simple Change Point Detection Algorithm — 88
  - 3.2.2 Change-Point Detection R Code — 89
- 3.3 Gap Filling with SSA — 92
- 3.4 Denoising by SSA — 96
  - 3.4.1 Filter Based Correlation Coefficients — 97

4 More on Filtering and Forecasting by SSA — 103
- 4.1 Introduction — 103
- 4.2 Filtering Coefficients — 104
- 4.3 Forecast Equation — 107
  - 4.3.1 Recurrent SSA Forecast Equation — 107
  - 4.3.2 Vector SSA Forecast Equation — 108
- 4.4 Different Window Length for Forecasting and Reconstruction — 111
- 4.5 Outlier in SSA — 112

Appendix A: A Short Introduction to R	117
Appendix B: Theoretical Explanations	137
Index	147

LIST OF FIGURES

Fig. 1.1	Quarterly US energy consumption time series (1973Q1–2015Q3)	12
Fig. 1.2	An approximations for the US energy consumption series using the first eigenvalue	13
Fig. 1.3	Plot of $w_j^{L,N}$ with respect to L, j for $N = 21$	19
Fig. 1.4	Matrix of w-correlations for the 24 reconstructed components of the energy series	20
Fig. 1.5	A realization of the simulated series	21
Fig. 1.6	Logarithms of the 200 simulated series eigenvalues	24
Fig. 1.7	Logarithms of the 24 singular values of the energy series	25
Fig. 1.8	Paired eigenfunctions 1–10 for the energy series	27
Fig. 1.9	Forecasts from Examples 1.6 and 1.7	32
Fig. 1.10	Comparing the last column of the approximated trajectory matrix, before and after diagonal averaging, for the US energy data	33
Fig. 1.11	Sensitivity analysis of RMSE of forecasts in US energy data	38
Fig. 1.12	Quarterly UK gas consumption time series over the period 1960Q1–1986Q4	41
Fig. 1.13	First nine Eigenfunctions for UK gas time series with $L = 46$	42
Fig. 1.14	W-correlations among pair components for UK gas time series with $L = 46$	43
Fig. 1.15	Comparison of forecasts by paris (i) ($L = 46, r = 7$) and (ii) ($L = 39, r = 11$) for UK gas consumption series; solid line with circle points show the original time series, dashed lines with triangle and square symbols show the forecasts by (i) and (ii), respectively	44

Fig. 1.16	Monthly UK government security yield time series over the period Jan. 1985–Dec. 2015	45
Fig. 1.17	First nine Eigenfunctions for UK government security yield time series with $L = 72$	46
Fig. 1.18	W-correlations among pair components for UK government security yield time series with $L = 72$	47
Fig. 1.19	Comparison of forecasts by user $(L = 72, r = 13)$ and automated choices $(L = 96, r = 3)$ for UK government security yield series	48
Fig. 2.1	Monthly number of US passengers, domestically and internationally, sample Oct. 2002–Oct. 2015	83
Fig. 2.2	Plot of Singular values of the trajectory matrix with Monthly number of US passengers, time series with $L = 72$	84
Fig. 2.3	Comparison of two forecasting scenarios with real observations	85
Fig. 3.1	Initial data (left side) and change-point detection statistic D_t (right side) in Example 3.1	91
Fig. 3.2	Fitting the trend (left side) and the change-point detection statistic D_t (right side) in Example 3.2	92
Fig. 3.3	Logarithm of the singular values of the trajectory matrix in Example 3.3, when $L = 9$	93
Fig. 3.4	Real interest rate in Japan during 1961–2014	95
Fig. 3.5	Scree plot of singular values of the HMSSA trajectory matrix, when $L = 13$ in Example 3.4	96
Fig. 3.6	Real signal (right side) and noisy data (left side) in Example 3.5	99
Fig. 3.7	Scatterplot for number of technicians and export percent in Example 3.6	100
Fig. 4.1	Vector SSA forecasting	109
Fig. 4.2	Plot of the first 10 paired eigenvectors for death series	114
Fig. 4.3	Plot of the death series and fitted curve	115
Fig. A.1	Some of the computational functions available in R	123
Fig. A.2	Several graphical functions	124
Fig. A.3	Examples of several graphical functions	124
Fig. A.4	Different point characters (pch) for plot function	125
Fig. A.5	Using the mfrow function	126
Fig. A.6	Using the layout function	127
Fig. A.7	Using the mfrow function	128
Fig. A.8	Examples of several graphical functions	134

LIST OF TABLES

Table 1.1	Summary of SSA and PCA processes	14
Table 1.2	The value of w-correlation for different values of L, $N = 200$	22
Table 1.3	Number of observations used in SSA for different L and $N = 20, 25, 30$	23
Table 2.1	Similarities and dissimilarities between the VMSSA and HMSSA recurrent forecasting algorithms	69
Table A.1	A discrete distribution	132

CHAPTER 1

Univariate Singular Spectrum Analysis

Abstract A concise description of univariate Singular Spectrum Analysis (SSA) is presented in this chapter. A step-by-step guide for performing filtering, forecasting as well as forecasting interval using univariate SSA and associated R codes is also provided. After reading this chapter, the reader will be able to select two basic, but very important, choices of SSA: window length and number of singular values. The similarity and dissimilarity between SSA and principal component analysis (PCA) is also briefly deliberated.

Keywords Univariate SSA · Window length · Singular values
Reconstruction · Forecasting

1.1 Introduction

There are several different methods for analysing time series all of which have sensible applications in one or more areas. Many of these methods are largely parametric, for example, requiring linearity or nonlinearity of a particular form. An alternative approach uses non-parametric techniques that are neutral with respect to problematic areas of specification, such as linearity, stationarity and normality. As a result, such techniques can provide a reliable and often better means of analysing time series data. Singular Spectrum Analysis (SSA) is a relatively new non-parametric method that has proved its capability in many different time series applications ranging from economics to physics. For the history of SSA, see Broomhead et al. (1987), and Broomhead and King (1986a, b). SSA has subsequently been developed in several ways including multivariate SSA (Hassani and Mahmoudvand 2013), SSA based on minimum variance (Hassani 2010) and SSA based on perturbation (Hassani et al. 2011b) (for more information, see Sanei and Hassani (2016)).

The increased application of SSA is further influenced by the following. Firstly, the emergence of Big Data may increase noise in time series, which in turn results in a distortion of the signal, thereby hindering the overall forecasting process. Secondly, volatile economic conditions ensure that time series (in most cases) are no longer stationary in mean and variance, especially following recessions which have left behind structural breaks. This in turn results in a violation of the parametric assumptions of stationarity and prompts data transformations when adopting classical time series methods. Such data transformations result in a loss of information and by relying on a technique such as SSA, which is not bound by any assumptions, users can overcome the restrictions imposed by parametric models in relation to the structure of the data. It is also noteworthy that recently it has been shown that SSA can provide accurate forecasts before, during and after recessions. Thirdly, SSA can be extremely useful as it enables the user to decompose a time series and extract components such as the trend, seasonal components and cyclical components (Sanei and Hassani 2016), which can then be used for enhancing the understanding of the underlying dynamics of a given time series. Fourthly, SSA is also known for its ability to deal with short time series where classical methods fail due to a lack of observations (Hassani and Thomakos 2010).

A common problem in economics is that most of the times series we study contain many components such as trend, harmonic and cyclical comments, and irregularities. Trend extraction or filtering are difficult even if we assume there is a time series with additive components. In general, as in SSA too, the trend of a time series is considered as a smooth additive component that contains information about the general tendency of the series. The most frequently used approaches for trend extraction are, for instance, simple linear regression model, moving average filtering, Tramo-Seats, X-11, X-12, and the most common one, the Hodrick-Prescott (HP) filter. To apply each method, one needs to consider model's specification or parameters. Generally, one can classify trend extraction approaches into tow main categorizes; the Model-Based approach, and non-parametric approaches including SSA. The Model-Based approach assumes the specification of a stochastic time series model for the trend, which is usually either an ARIMA model or a state space model. On the other hand, the non-parametric filtering methods (i.e. the Henderson, and Hodrick-Prescott filters) do not require specification of a model; they are quite easy to apply and are used in all applied areas of time series analysis. However, there are a few disadvantages of using HP filter; (i) "the HP filter produces series with spurious dynamic relations that have no basis in the underlying data-generating process; (ii) a one-sided version of the filter reduces but does not eliminate spurious predictability and moreover produces series that do not have the properties sought by most potential users of the HP filter" Hamilton (2017).

Two main important applications of SSA are filtering and smoothing, and forecasting, which will be discussed in the following sections.

1.2 Filtering and Smoothing

The SSA technique decomposes the original time series into the sum of a small number of interpretable components, such as a slowly varying trend, oscillatory components and noise. The basic SSA method consists of two complementary stages: decomposition and reconstruction, of which each stage includes two separate steps. At the first stage the series is decomposed and, in the second stage, the filtered series is reconstructed; the reconstructed series is then used for forecasting new data points. A short description of the SSA technique is given below (for more details, see Hassani et al. 2012).

Stage I. Decomposition

We consider a stochastic process Y generating a sequence comprising N random variables: $Y_N \equiv \{Y_t\} \equiv \{Y_t\}_{t=1}^{N}$. The sequence is ordered in time. In practice, we deal with realizations, or outcomes, from this process which we index by $t = 1, \ldots, N$, and distinguish them from the underlying random variables by using lower case y, that is $Y_N = (y_1, \ldots, y_N)$.

1st Step: Embedding. Embedding can be considered as a mapping which transfers a one-dimensional time series $Y_N = (y_1, \ldots, y_N)$ into a multi-dimensional series $X_1, \ldots, X_K$ with vectors $X_i = (y_i, \ldots, y_{i+L-1})^T \in \mathbf{R}^L$, where L is the *window length* (see Sect. 1.4.1), and $2 \leq L \leq N/2$ and $K \equiv N - L + 1$. The single input at this stage is the SSA choice of L. The result of this step is the trajectory matrix:

$$\mathbf{X} = [X_1, \ldots, X_K] = (x_{ij})_{i,j=1}^{L,K}$$

$$= \begin{pmatrix} x_{11} & x_{12} & \cdots & x_{1K} \\ x_{21} & x_{22} & \cdots & x_{2K} \\ \vdots & \vdots & \ddots & \vdots \\ x_{L1} & x_{L2} & \cdots & x_{LK} \end{pmatrix} \equiv \begin{pmatrix} y_1 & y_2 & \cdots & y_K \\ y_2 & y_3 & \cdots & y_{K+1} \\ \vdots & \vdots & \ddots & \vdots \\ y_L & y_{L+1} & \cdots & y_N \end{pmatrix}$$

(1.1)

Note that the output from the embedding step is the trajectory matrix $\mathbf{X}$, which is a Hankel matrix. This means that all the elements along the diagonal $i + j = const$ are equal, for example, $x_{12} = x_{21} = y_2$. Note also that, the first column of $\mathbf{X}$ includes the observations 1 to L of the time series, the second column corresponds to observations 2 to $L+1$ and so on. One preference in preparing SSA is the use of matrices rather than vectors. Moreover, the majority of signal processing techniques can be seen as applied linear algebra and thus we are able to benefit accordingly. If a time series Y is defined in R (for more information about defining new time series, see Appendix A), then, the following R code will produce the Hankel matrix $\mathbf{X}$.

> **Example 1.1: Constructing a Trajectory matrix in R: deterministic series**

Let us begin by generating an illustrative Hankel matrix $\mathbf{X}$ from a determined time series Y given by:

```
Y<-1:15
```

The next step is to select a value for L which is the input at the embedding step.

```
L<-7 #
K<-length(Y)-L+1
X<-outer((1:L),(1:K),function(x,y) Y[(x+y-1)])
```

Finally, in order to see the Hankel matrix typing in X results in the following output:

	[,1]	[,2]	[,3]	[,4]	[,5]	[,6]	[,7]	[,8]	[,9]
[1,]	1	2	3	4	5	6	7	8	9
[2,]	2	3	4	5	6	7	8	9	10
[3,]	3	4	5	6	7	8	9	10	11

```
[4,]   4   5   6   7   8   9  10  11  12
[5,]   5   6   7   8   9  10  11  12  13
[6,]   6   7   8   9  10  11  12  13  14
[7,]   7   8   9  10  11  12  13  14  15
```

> **Example 1.2 Constructing the Trajectory matrix in R: random series**

In this example we generate a time series from a random uniform distribution. The runif function below generates random deviates from a uniform distribution with $N = 7$ between a minimum of 0 and maximum of 1, with L=4.

```
Y1=round(runif(7,0,1),1)
Y1<-c(0.8,0.5,0.9,0.4,0.7,0.1,0.6)
L<-4
K<-length(Y1)-L+1
X<-outer((1:L),(1:K),function(x,y) Y1[(x+y-1)])
X
     [,1] [,2] [,3] [,4]
[1,]  0.8  0.5  0.9  0.4
[2,]  0.5  0.9  0.4  0.7
[3,]  0.9  0.4  0.7  0.1
[4,]  0.4  0.7  0.1  0.6
```

2nd Step: Singular Value Decomposition (SVD). In this step, the SVD of **X** is performed. Denote by $\lambda_1, \ldots, \lambda_L$ the eigenvalues of $\mathbf{XX}^T$ arranged in decreasing order ($\lambda_1 \geq \cdots \lambda_L \geq 0$) and by $U_1, \ldots, U_L$ the corresponding left eigenvectors. The SVD of **X** can be written as $\mathbf{X} = \mathbf{X}_1 + \cdots + \mathbf{X}_L$, where $\mathbf{X}_i = \sqrt{\lambda_i} U_i V_i^T$ and $V_i = \mathbf{X}^T U_i / \sqrt{\lambda_i}$ (if $\lambda_i = 0$, then set $\mathbf{X}_i = 0$). The $\mathbf{X}_i$ matrices are referred to in SSA as elementary matrices. Here, the $\sqrt{\lambda_i}$ are referred to as the singular values of **X**, and the collection $\{\sqrt{\lambda_1}, \sqrt{\lambda_2}, \ldots, \sqrt{\lambda_L}\}$ is called the spectrum. The name "Singular Spectrum Analysis" comes from this property of the technique and is a vital component as the SSA process is concentrated around obtaining and analysing this spectrum of singular values, to identify and distinguish between the signal and noise in a given time series.

The function svd in R computes the SVD of a matrix and the following codes give eigenvalues and eigenvectors of the Hankel matrix **X**.
In order to obtain the SVD of **X** use the following code:

```
SVD<-svd(X)
```

The singular values of **X**, that is the $\sqrt{\lambda_i}$, for any given time series can be extracted using the code:

```
lambda<-sqrt(SVD$d)
```

Likewise, the left and right eigenvectors for a given time series can be extracted as follows:

```
U<-SVD$u
V<-SVD$v
```

In these codes, U and V are $L \times L$ and $L \times K$ matrices and their columns are $U_1, \ldots, U_L$ and $V_1, \ldots, V_K$, respectively. Moreover, lambda contains the singular values $\sqrt{\lambda_1}, \ldots, \sqrt{\lambda_L}$ and also defines a diagonal matrix, Λ, with the singular values on its main diagonal. In this way, there is the following equality by SVD:

$$\mathbf{X} = \mathbf{U}\Lambda\mathbf{V}^T, \qquad (1.2)$$

where Λ denotes the diagonal matrix with diagonal entities $\sqrt{\lambda_1}, \ldots, \sqrt{\lambda_L}$. This equality can be checked by the following code in R.

```
U%*%Lambda%*%t(V)
```

To identify the $\mathbf{X}_i$ matrices in R, use the following code:

```
Xi<-lambda[i]*U[,i]%*%t(V[,i])
```

Example 1.3 The Singular Value Decomposition

Applying the above codes, the first component X1 of Example 1.1 can be obtained as below (Note that some of the R conventions are explained in the Appendix A):

```
SVD<-svd(X)
lambda<-SVD$d
Lambda<-diag(lambda)
U<-SVD$u
V<-SVD$v
X1<-lambda[1]*U[,1]%*%t(V[,1])
round(X1,2)
     [,1] [,2] [,3] [,4] [,5] [,6] [,7] [,8] [,9]
[1,] 2.79 3.42 4.04 4.67 5.29 5.92 6.55 7.17 7.80
[2,] 3.27 4.01 4.74 5.48 6.21 6.94 7.68 8.41 9.15
[3,] 3.75 4.60 5.44 6.28 7.12 7.97 8.81 9.65 10.49
[4,] 4.24 5.19 6.14 7.09 8.04 8.99 9.94 10.89 11.84
```

```
[5,] 4.72 5.78 6.84 7.89  8.95 10.01 11.07 12.13 13.19
[6,] 5.20 6.37 7.53 8.70  9.87 11.04 12.20 13.37 14.54
[7,] 5.68 6.96 8.23 9.51 10.78 12.06 13.33 14.61 15.88
```

The two steps of embedding and SVD, complete the decomposition stage of SSA. We move now to stage II, that of reconstruction.

Stage II. Reconstruction

There are two steps in the reconstruction of matrices, namely grouping and diagonal averaging, which comprise the second stage of SSA.
1st Step: Grouping. The grouping step corresponds to splitting the elementary matrices into several groups and summing the matrices within each group. A square matrix is called an elementary matrix if it can be obtained from an identity matrix using a single elementary row operation. The aim here is to enable the signal and noise to be distinguished.

Splitting the set of indices $\{1,\ldots,L\}$ into disjoint subsets $I_1,\ldots,I_m$ corresponds to the representation $\mathbf{X} \equiv \mathbf{X}_{I_1} + \cdots + \mathbf{X}_{I_m}$, where

$$\mathbf{X}_{I_j} = \sum_{\ell \in I_j} \mathbf{X}_\ell, \quad j=1\ldots,m.$$

The procedure of choosing the sets $I_1,\ldots,I_m$ is called grouping. Below is an example of the R code that performs this grouping.

Example 1.4 Grouping

As an illustration assume that the matrix $\mathbf{X}$ from Example 1.2 is reconstructed using components 2 and 3, where components refer to eigenvalues. Then, $I_1 = \{2,3\}$ results in:

```
I1<-c(2,3)
p<-length(I1)
XI1<-U[,I1]%*%matrix(Lambda[I1,I1],p,p)%*%t(V[,I1])
XI1
      [,1]  [,2]  [,3]  [,4]  [,5]  [,6]  [,7]  [,8]  [,9]
[1,] -1.79 -1.42 -1.04 -0.67 -0.29  0.08  0.45  0.83  1.20
[2,] -1.27 -1.01 -0.74 -0.48 -0.21  0.06  0.32  0.59  0.85
[3,] -0.75 -0.60 -0.44 -0.28 -0.12  0.03  0.19  0.35  0.51
[4,] -0.24 -0.19 -0.14 -0.09 -0.04  0.01  0.06  0.11  0.16
[5,]  0.28  0.22  0.16  0.11  0.05 -0.01 -0.07 -0.13 -0.19
[6,]  0.80  0.63  0.47  0.30  0.13 -0.04 -0.20 -0.37 -0.54
[7,]  1.32  1.04  0.77  0.49  0.22 -0.06 -0.33 -0.61 -0.88
```

As you will see in the following sections, at the grouping step we have the option of analysing the periodogram, scatterplot of right eigenvectors or the eigenvalue functions graph to differentiate between noise and signal. Once we have selected the eigenvalues corresponding to the noise and signal, we can then evaluate the effectiveness of this separability via the weighted correlation (w-correlation) statistic. The w-correlation measures the dependence between any two time series (here, e.g. consider $Y_N^{(1)}$ and $Y_N^{(2)}$ each reconstructed using the eigenvalues in $\mathbf{X}_1$ and $\mathbf{X}_2$, respectively) and if the separability is sound then the two time series will report a w-correlation of zero. In contrast, if the w-correlation between the reconstructed components is large, then this indicates that the components should be considered as one group.

2nd Step: Diagonal averaging. The purpose of diagonal averaging is to transform a matrix to the form of a Hankel matrix, which can be subsequently converted to a time series. If z_{ij} stands for an element of a matrix $\mathbf{Z}$, then the k-th term of the resulting series is obtained by averaging z_{ij} over all i, j, such that $i + j = k + 1$. By performing the diagonal averaging of all matrix components of $\mathbf{X}_{I_j}$ in the expansion of $\mathbf{X}$ in the grouping step, another expansion is obtained: $\mathbf{X} = \widetilde{\mathbf{X}}_{I_1} + \cdots + \widetilde{\mathbf{X}}_{I_m}$, where $\widetilde{\mathbf{X}}_{I_j}$ is the diagonalized version of the matrix $\mathbf{X}_{I_j}$. This is equivalent to the decomposition of the initial series $Y_N = (y_1, \ldots, y_N)$ into a sum of m series; $y_t = \sum_{j=1}^{m} \widetilde{y}_t^{(j)}$, where $\widetilde{Y}_N^{(j)} = (\widetilde{y}_1^{(j)}, \ldots, \widetilde{y}_N^{(j)})$ corresponds to the matrix $\widetilde{\mathbf{X}}_{I_j}$.

It is worth mentioning, that if $\widetilde{x}_{r,s}$ is the rsth entry of the matrix $\widetilde{\mathbf{X}}_{I_j}$, then applying the diagonal averaging formula it follows that:

$$\widetilde{y}_t^{(j)} = \frac{1}{s_2 - s_1 + 1} \sum_{i=s_1}^{s_2} \widetilde{x}_{i,t+1-i}, \tag{1.3}$$

where $s_1 = \max\{1, t - N + L\}$, $s_2 = \min\{L, t\}$.

> **Example 1.5 Diagonal Averaging**

Let XI be a matrix that is obtained with $I_1 = \{1\}$ in the grouping step of Example 1.1, then the following code will produce an approximation of the original series. We term the output from the code below as an 'approximation' as the reconstructed series is obtained with only the first eigenvalue.

```
D<-NULL
N<-length(Y)
for(t in 1:N){
s1<-max(1,(t-N+L))
s2<-min(L,t)
place<-(s1:s2)+L*(((t+1-s1):(t+1-s2))-1)
D[t]<-mean(XI[place])}
round(D,2)
[1]   2.79  3.34  3.93  4.56  5.22  5.92  6.66  7.61
[9]   8.56  9.69 10.86 12.06 13.30 14.57 15.88
```

The four functions to perform steps 1 and 2 of the first stage and steps 1 and 2 of the second stage of SSA are given below. They are called UniHankel(), SVD(), Group() and DiagAver(), respectively.

We begin by performing the Hankelization.

Program 1.1 Hankelization R code

```
UniHankel<-function(Y,L){
  k<-length(Y)-L+1
  outer((1:L),(1:k),function(x,y) Y[(x+y-1)])
}
```

Then obtain the SVD.

Program 1.2 SVD R code

```
SVD<-function(Y,L){
  X<-UniHankel(Y,L)
  svd(X)
}
```

Followed by the grouping process.

Program 1.3 Grouping R code

```
Group<-function(Y,L,groups){
  I<-groups;p<-length(I)
  SVD<-SVD(Y,L)
  LambdaI<-matrix(diag(SVD$d)[I,I],p,p)
  SVD$u[,I]%*%LambdaI%*%t(SVD$v[,I])
}
```

Finally, perform Diagonal Averaging so that the matrix can be converted into a time series.

Program 1.4 Diagonal Averaging R code

```
DiagAver<-function(X){
 L<-nrow(X);k<-ncol(X);N<-k+L-1
 D<-NULL
 for(j in 1:N){
   s1<-max(1,(j-N+L))
   s2<-min(L,j)
   place<-(s1:s2)+L*(((j+1-s1):(j+1-s2))-1)
   D[j]<-mean(X[place])
   }
 D
}
```

Applying these functions, it is possible to write a general function to calculate the components of a time series by SSA. In this case the function is called `SSA.Rec` and is defined as follows:

Program 1.5 SSA Reconstruction R code

```
SSA.Rec<-function(Y,L,groups){
  N<-length(Y)
  I<-groups;p<-length(I)
  XI<-Group(Y,L,groups)
  Approx<-DiagAver(XI)
  Resid<-Y-Approx
 list(Approximation=Approx,Residual=Resid)
}
```

Note that to execute function `SSA.Rec`, functions `UniHankel`, `SVD`, `Group` and `DiagAver` must be defined in R. It is advisable to write the whole functions sequentially in a text file and then copy the content of that file into the command line of R to apply `SSA.Rec`.

Example 1.6 Reconstruction in R

Here `Approximation` shows 15 data points of the time series in Example 1.1 that are initially decomposed using $L = 7$ and reconstructed using the first eigenvalue alone. The `Residual` is the difference between the actual values and the approximated values.

```
 SSA.Rec(1:15,7,c(1))
$Approximation
 [1]    2.789869    3.343960    3.934112    4.560325
 [5]    5.222600    5.920935    6.655331    7.606092
 [9]    8.556854    9.687920   10.855047   12.058235
[13]   13.297484   14.572794   15.884164

$Residual
 [1]  -1.78986857  -1.34395995  -0.93411226
 [4]  -0.56032548  -0.22259963   0.07906529
 [7]   0.34466930   0.39390777   0.44314624
[10]   0.31208010   0.14495304  -0.05823495
[13]  -0.29748385  -0.57279368  -0.88416443
```

> **Example 1.7 Running basic SSA steps in a Real data**

To illustrate the 4 steps in SSA, we consider an example using US energy consumption data. The series shown in Fig. 1.1 is of the quarterly energy consumption in the USA between 1973Q1 and 2015Q3. These data can be accessed via http://www.eia.gov/totalenergy/data/monthly/#consumption.

All the results and figures in this example are obtained by means the R functions defined above. There is a variety of options for loading data into R[1] and in this instance we rely on the easiest and most basic method of data importation and begin by 'scanning' the energy series into the R platform using the approach shown below. When using this approach, users simply need to copy their data and paste it into R. The data for this example, are read in and saved as `energy` in R and are shown below:

`energy<-scan()`

By calling 'energy' users can then view the observations on the time series. Here, we demonstrate some part of the data in order to save space.

```
energy
     .
     .
     .
[145] 6613.619 4397.948 4839.245 5257.553 6818.974 4389.818
[151] 5237.253 5398.310 6684.719 4438.833 5235.077 5044.533
```

---

[1] http://www.statmethods.net/input/importingdata.html.

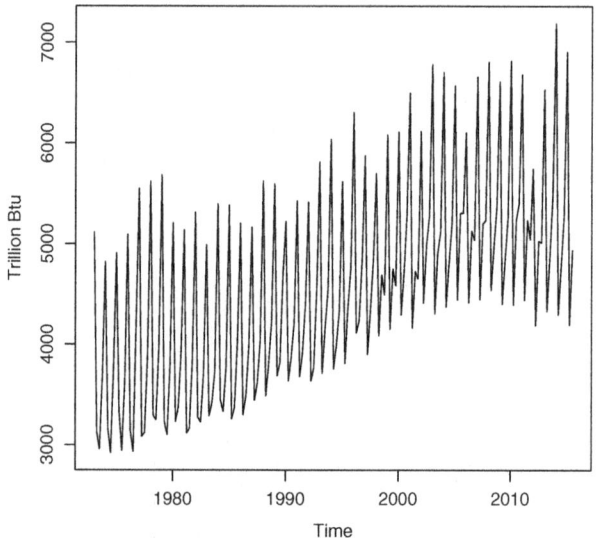

Fig. 1.1 Quarterly US energy consumption time series (1973Q1–2015Q3)

```
[157] 5743.924 4187.127 5028.028 5012.615 6534.639 4327.458
[163] 4869.267 5466.510 7190.220 4293.271 4778.453 5280.111
[169] 6910.090 4194.068 4939.494
```

The following code will produce the plot of this series as in Fig. 1.1:

```
plot(ts(energy,frequency=4, start=c(1973,1)),xlab= "Time",
    ylab= "Trillion Btu")
```

From Fig. 1.1 observe that the energy series portrays seasonality and, therefore, we now seek to extract the related harmonic components in the steps which follow.

An approximation of the energy series, with the first eigenvalue is depicted in Fig. 1.2. These plots can be obtained by the following codes:

```
Approx1<-SSA.Rec(energy,24,c(1))$Approximation
Data<-cbind(energy,Approx1)
Energy<-ts(Data,frequency=4,start=c(1973,1))
plot.ts(Energy[,1],xlab= "Time", ylab= "Trillion Btu")
legend("topleft",horiz=FALSE,bty = "n", lty=c(1,2),
,c("real","EF 1"),lwd=c(1,2))
lines(Energy[,2],lty=2,lwd=2)
```

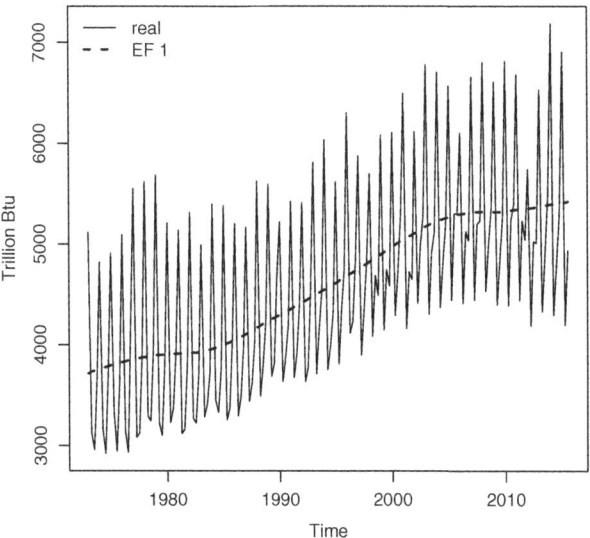

Fig. 1.2 An approximations for the US energy consumption series using the first eigenvalue

1.3 COMPARING SSA AND PCA

Consider a data matrix as below:

$$\mathbf{X} = \begin{pmatrix} x_{11} & x_{12} & \cdots & x_{1p} \\ x_{21} & x_{22} & \cdots & x_{2p} \\ \vdots & \vdots & \ddots & \vdots \\ x_{n1} & x_{n2} & \cdots & x_{np} \end{pmatrix}, \quad (1.4)$$

where each of the n rows represents a sample, and each of the p columns gives a particular kind of feature. In order to use PCA, there are no restrictions on observations x_{ij}. But, n and p are fixed and p must be greater than 2. In contrast, univariate SSA start with a univariate vector $(y_1, \ldots, y_N)$ and produce data matrix:

Table 1.1 Summary of SSA and PCA processes

SSA Process	PCA Process
(i) Get univariate series $Y_N = (y_1, \ldots, y_N)$	
(ii) Construct trajectory matrix $\mathbf{X}$ of dimension $L \times K$	(i) Get data matrix $\mathbf{X}$ of dimension $n \times p$
(iii) Decompose trajectory matrix by SVD as $\mathbf{X} = \mathbf{U}\Lambda\mathbf{V}'$	(ii) Decompose data matrix by SVD as $\mathbf{X} = \mathbf{U}\Lambda\mathbf{V}'$
(iv) Do grouping and compute $\tilde{\mathbf{X}}_{I_j} = \sum_{\ell \in I_j} U_\ell U_\ell' \mathbf{X}$ for group j	(iii) Compute PC's as $\mathbf{Z} = \mathbf{XV} = \mathbf{U}\Lambda$ of dimension $n \times q$
(v) Compute diagonal averaging and obtain $\tilde{Y}_N^{(j)}$ for group j	(iv) Choose first r $(r < q)$ columns of $\mathbf{Z}$ as the main PC's

$$\mathbf{X} = \begin{pmatrix} y_1 & y_2 & \cdots & y_K \\ y_2 & y_3 & \cdots & y_{K+1} \\ \vdots & \vdots & \ddots & \vdots \\ y_L & y_{L+1} & \cdots & y_N \end{pmatrix}, \quad (1.5)$$

where $k = N - L + 1$. This matrix is a Hankel matrix, means that there are restriction on the elements of the matrix. Unlike PCA where the number of rows and columns are fixed, the rows number in SSA, L, can be adjusted between 2 and $N/2$. This means that the subspaces in PCA is limited, whereas the subspaces in SSA can be adjusted by varying the window length L. However, both SSA and PCA use SVD in their algorithms.

Table 1.1 shows a summary of PCA and SSA processes. As it indicated PCA takes a data matrix as input, perform an operation and then output a resulting matrix. In contrast, SSA take a univariate vector, construct a trajectory matrix and then use a PCA process on this matrix and finally transform the results to a univariate vector.

The above comparison shows that we couldn't use PCA for univariate time series directly. Indeed, SSA uses PCA to analyze univariate time series.

1.4 Choosing Parameters in SSA

The complete procedure of the SSA technique depends upon two basic, but very important, choices that influence the results, namely:

(i) the window length L,
(ii) the way of grouping, include the number of groups and the member(s) of each groups. In a simple case with only two groups, one for the signal and the other for the noise, grouping means finding a cut-off point r so that the first r components will be used to reconstruct the signal and the remaining for noise.

Choosing improper parameters yields an incomplete reconstruction and misleading results in forecasting. In spite of the importance of choosing parameters, no theoretical solution has been yet proposed to this problem. Of course, there are worthwhile efforts and various techniques for selecting the appropriate value of L (see, e.g., Golyandina et al. (2001), Hassani et al. (2012), Golyandina (2010), Hassani et al. (2011a), Mahmoudvand and Zokaei (2012) and Mahmoudvand et al. (2013)). This is an area of current research by the authors and others, see, for example, Khan and Poskitt (2013a, b).

1.4.1 Window Length

The window length is the sole SSA choice in the decomposition stage of SSA. The choice of parameter L depends on the data and the purpose of the analysis where the danger is that some choices of L lead to an inferior decomposition.

On the selection of L, some discussion is given by spscitetEls who remarked that choosing $L = N/4$ is a common practice. Golyandina et al. (2001) recommend that L should be 'large', but not larger than $N/2$. Large values of L allow longer period oscillations to be resolved; however, choosing L too large leaves too few observations from which to estimate the covariance matrix of the L variables. It should be noted that variations in L may influence the separability feature of the SSA technique through the orthogonality and the closeness of the singular values. There are some methods for selecting L, for example, the weighted correlation between the signal and noise component has been proposed in Golyandina et al. (2001) to determine a suitable value of L in terms of separability. A detailed discussion on this topic can be found in Golyandina et al. (2001) (Sect. 1.6) and Golyandina (2010).

There are versions of univariate SSA where the window length L is chosen automatically (see, e.g. Alvarez-Meza et al. 2013). An information theoretic analysis of the signal-noise separation problem in SSA has also been proposed in Khan and Poskitt (2010). Given the lack of a firm theoretical background for selecting L, here we seek to provide a rationale for a choice of L in the interval $[2, N/2]$, with a time series of length N. More details can be found in Hassani et al. (2011a, 2012), Mahmoudvand and Zokaei (2012) and Mahmoudvand et al. (2013). There are several criteria for determining window length and we categorize them into two groups as follows:

1. Criteria that consider different features of SSA without regard to the type of data (general-based criteria).
2. Criteria that consider the object of analysis and depend on the type of data (problem-based criteria).

Some recommendations will be given for choosing the window length according to these criteria. In order to find the optimal value of window length L, the structure of the Hankel matrix (which is the first output of the first step of SSA) and the error of reconstruction step (using the root mean squared error) are investigated. Throughout this chapter, we consider a time series $Y_N = S_N + \varepsilon_N$ of length N, where S_N is the component of interest (usually called the signal) and ε_N is the noise component (which can be random noise or a deterministic component). The aim of the SSA reconstruction stage is to find an estimate for the signal component S_N, $\hat{S}_N$. In the ideal situation the noise component ε_N is completely removed; i.e, $S_N = \hat{S}_N$, however, in practical situations it is not possible to reconstruct S_N perfectly and in this chapter we explain how to find $\hat{S}_N$ close to S_N with respect to different criteria.

Rank of the Trajectory matrix

Considering the trajectory matrix $\mathbf{X}$ of dimension $L \times K$, whose entities are defined by Eq. (1.1), we can say that the maximum number of components that we can obtain in decomposing the corresponding time series is equal to $L \times K$. This number is the maximum rank of the trajectory matrix and it is easy to see that the maximum rank is attainable when $L = L_{\max}$, where

$$L_{\max} = \begin{cases} \dfrac{N+1}{2} & \text{if N is odd,} \\ \dfrac{N}{2}, \dfrac{N}{2}+1 & \text{if N is even.} \end{cases}$$

Note that $L_{\max}$ is the median of $\{1, \ldots, N\}$ (see Theorem B.1).

The Lag-Covariance Matrices

We consider the behavior of the matrix $\mathbf{XX}^T/K$, which can be viewed as the lag-covariance matrix. Note that K indicates the number of lagged vectors as in classical multivariate analysis. For a fixed value of L, denote by $T_{\mathbf{X}}^{L,N}$ the trace of the matrix $\mathbf{XX}^T$, $tr(\mathbf{XX}^T)$, we can show that:

$$T_{\mathbf{X}}^{L,N} = T_{\mathbf{X}}^{K,N} = \sum_{j=1}^{N} w_j^{L,N} y_j^2 = \sum_{j=1}^{L} \lambda_j, \qquad (1.6)$$

where $w_j^{L,N} = \min\{\min\{L, K\}, j, N-j+1\} = w_j^{K,N}$ and $K = N - L + 1$, (see Appendix B, Theorem B.3). In addition, Theorem B.4 shows that:

$$\max_{L \in \{2,\ldots,N-1\}} T_{\mathbf{X}}^{L,N} = T_{\mathbf{X}}^{\max,N}. \qquad (1.7)$$

Recall that SSA is a technique that uses several steps to convert a time series into a set of interpretable components. This transformation is defined in such a way that the first component has the largest possible variance (that, it accounts for as much of the variability in the series as possible), and each succeeding component in turn has the highest variance possible under the constraint that it is well separated from the preceding components. Equation (1.7) shows that $L_{\max}$ produces the highest total variability. For more information refer to Appendix B, Theorem B1.

Separability (w-correlation)

The main concept in studying SSA properties is separability, which characterizes how well different components can be separated from each other. The SSA decomposition of the series Y_N can only be successful if the resulting additive components of the series are approximately separable from each other. There are two types of separability, strong and weak, and here we only discuss weak separability as strong separability rarely holds. A brief description of strong separability is provided in Appendix B. The following quantity called the *w-correlation*, is a natural measure of the similarity between two series $Y_N^{(i)}$ and $Y_N^{(j)}$, see Golyandina et al. (2001):

$$\rho_{12}^{(w)} = \frac{\left(Y_N^{(i)}, Y_N^{(j)}\right)_w}{\sqrt{\left(Y_N^{(i)}, Y_N^{(i)}\right)_w}\sqrt{\left(Y_N^{(j)}, Y_N^{(j)}\right)_w}},$$

where $\left(Y_N^{(i)}, Y_N^{(j)}\right)_w = \sum_{p=1}^N w_p^{L,N} y_p^{(i)} y_p^{(j)}$, $(i, j = 1, 2)$. If the absolute value of the w-correlation is small, then the corresponding series are almost w-orthogonal, however, if it is large, then the two series are far from being w-orthogonal and are termed weakly separable. Hassani et al. (2011a) have shown that the minimum value of the w-correlations is attained at $L = L_{max}$, for a wide class of time series. For more information, see Hassani et al. (2011b) and Mahmoudvand et al. (2013), who provided several different examples from real data and simulated examples supporting this conclusion.

The R function for computing the w-correlation for two arbitrary vectors $Y1$ and $Y2$ is given below:

```
w.corr<-function(L,Y1,Y2){
N<-length(Y1)
w.corr<-NULL
w<-((N+1)-abs((N+1)/2-L)-abs((N+1)/2-1:N)-
+ abs(abs((N+1)/2-L)-abs((N+1)/2-1:N)))/2
sum((w*Y1)*Y2)/sqrt(sum(w*Y1^2)*sum(w*Y2^2))
}
```

In this code, we have used an alternative form of the weights $w_j^{L,N}$, which was obtained by Mahmoudvand and Zokaei (2012) as below:

$$w_j^{L,N} = \frac{N+1}{2} - \frac{\left|\frac{N+1}{2} - L\right|}{2} - \frac{\left|\frac{N+1}{2} - j\right|}{2} - \frac{\left|\left|\frac{N+1}{2} - L\right| - \left|\frac{N+1}{2} - j\right|\right|}{2}.$$

These weights have several interesting properties which are elaborated in Appendix B. Figure 1.3 shows two samples of the behaviour of $w_j^{L,N}$ with respect to j, indicating maximum weights $w_j^{L,N}$ are obtained when $L = L_{max}$.

In SSA terminology, $Y1$ and $Y2$ are the components that are provided after the grouping step. The R function to compute the w-correlation in this case is provided below:

Program 1.6 w-correlation R code

```
W.corr<-function(Yt,L,groups){
m<-length(groups);w.corr<-diag(m)
N<-length(Yt)
w<-((N+1)-abs((N+1)/2-L)-abs((N+1)/2-1:N)-
abs(abs((N+1)/2-L)- abs((N+1)/2-1:N)))/2
```

1 UNIVARIATE SINGULAR SPECTRUM ANALYSIS

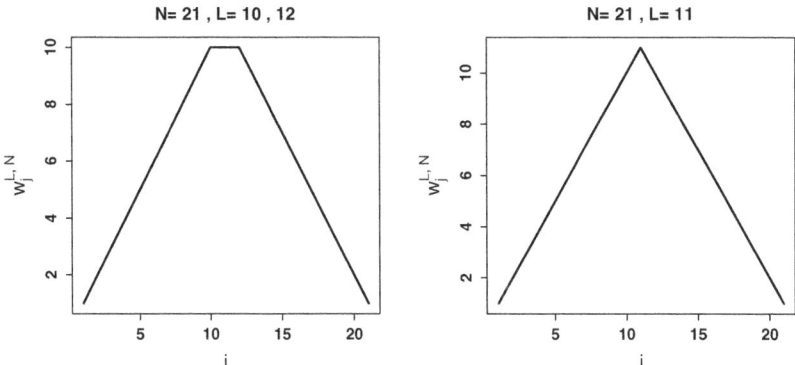

Fig. 1.3 Plot of $w_j^{L,N}$ with respect to L, j for $N = 21$

```
wcorr<-function(i,j){
Y1<-SSA.Rec(Yt,L,groups[[i]])$Approximation
Y2<-SSA.Rec(Yt,L,groups[[j]])$Approximation
sum(w*Y1*Y2)/sqrt(sum(w*Y1^2)*sum(w*Y2^2))}
for(i in 1:(m-1)){
for(j in (i+1):m){
w.corr[i,j]=w.corr[j,i]=wcorr(i,j)}}
rownames(w.corr)<-colnames(w.corr)<-groups
w.corr
}
```

Note that, the input argument groups in this function includes the groups and must be defined as a list. A sample of the results of this code for the US energy series is shown in Fig. 1.4. The w-correlation matrix is able to indicate how well SSA separates the signal from the noise. A w-correlation value which is close to zero indicates the components are w-orthogonal and therefore weakly separable. Large values of w-correlations between reconstructed components indicate that the components should possibly be gathered into one group and correspond to the same component in SSA decomposition. The w-correlation value between pure sine and cosine with equal frequencies, amplitudes and phases is very high and close to 1.

According to Fig. 1.4, the first five eigentriples construct the signal series and remaining components the noise series. It is clearly seen that splitting of all the eigentriples into two groups, from the first to the fifth and the rest, gives rise to a decomposition of trajectory matrix into two almost orthogonal blocks, with the first block corresponding to the signal and the second block corresponding to noise. Note also that the w-correlation value

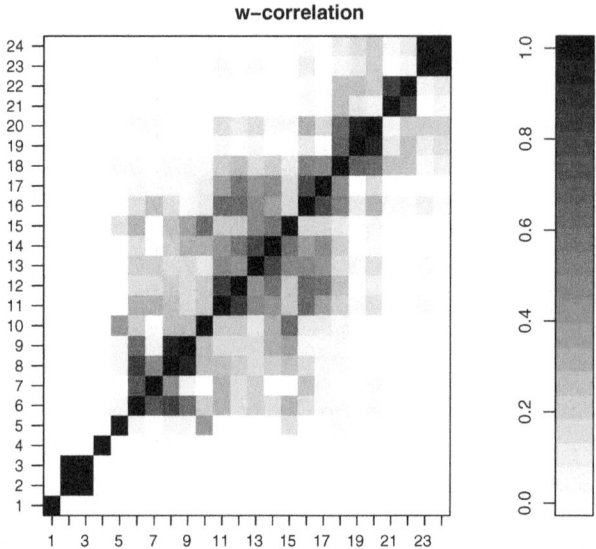

Fig. 1.4 Matrix of w-correlations for the 24 reconstructed components of the energy series

between reconstructed components by eigentriples 2 and 3 is very high and close to 1. These components are related to a harmonic component with specific frequency that can be detected with periodogram analysis.

```
m<-W.corr(energy,24,seq(1:24))
Plt.Img(m)
title(main="w-correlation")
```

Example 1.8 W-correlation in a simulated data

As a simulated example, consider the application of SSA for analysing the following series where, we have used a dummy variable to generate a systematic outlier:

$$y_t = 0.7\,sin(2\pi t/17) + 0.5\,sin(2\pi t/7) + 0.2\,sin(2\pi t/3) + u_t I_{u_t} + 0.5\,\varepsilon_t, \tag{1.8}$$

where ε_t is a white noise process. The random variable u_t is distributed uniformly between 0.5 and 1, and I_{u_t} is defined as:

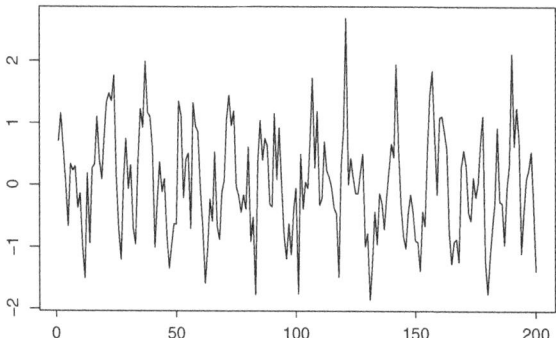

Fig. 1.5 A realization of the simulated series

$$I_{u_t} = \begin{cases} 1 \text{ if } t = 40\,k \quad (k = 1, \ldots, 4) \\ 0 \text{ otherwise} \end{cases} \quad (1.9)$$

The random variable $u_t I_{u_t}$ adds some peaks to the series systematically. The behaviour of these peaks is dynamic and completely different from the random variable ε_t, however, this part of the signal can be simply considered as the noise component, although it still contains useful information. Figure 1.5 displays the series, which is produced by the following R code:

```
ut<-runif(200,.5,1)
I.ut<-rep(0,200)
I.ut[c(40*(1:4))]<-1
et<-rnorm(200)
t<-1:200
yt<-0.7*sin(2*pi*t/17)+0.5*sin(2*pi*t/7)+
+ 0.2*sin(2*pi*t/3)+ut*I.ut+0.5*et
plot(t,yt,type="l",ylab=expression(y[t]))
```

Visual inspection of Fig. 1.5 indicates that the depicted series has a complicated structure and looks like a noise series. For more information about the structure of this series and the applicability SSA in this case, see Ghodsi et al. (2009).

According to the w-correlation criterion, if two reconstructed components have a zero w-correlation, then it means that these two components are separable. On the other hand, large values of the w-correlations between the reconstructed components indicate that the components should possibly be gathered into one group and correspond to the same component in

Table 1.2 The value of w-correlation for different values of L, $N = 200$

Window length	L=20	L=40	L=60	L=80	L=100
w-correlation	0.044	0.023	0.015	0.010	0.006

SSA decomposition. Table 1.2 gives the results obtained from the simulated series, which show that the minimum value of w-correlation is obtained for the maximum value of the window length $L = N/2$ confirming the theoretical results in the previous section.

The window length L is the only parameter in the decomposition stage and the theoretical and empirical guidance suggests that L should be large, but not greater than $N/2$. We therefore take $L = 100$ (in this case $N = 200$) and, based on this window length, and on the SVD of the trajectory matrix, there are 100 eigentriples, ordered by their contributions (shares) into the decomposition stage. The leading eigentriple describes the general tendency of the series. Since in most practical cases the eigentriples with small shares are related to the noise component of the series, the important practical step is to identify the set of leading eigentriples.

Number of Entities of the Trajectory Matrix

The aim of SSA is to decompose the original series into several components. In this context, it is clear that large values of L give more components and may give more chance for separability to hold. On the other hand, K can be considered the number of L-variate samples from original time series and thus large values of K can be considered as the better choice from a statistical point of view. These two aims can be obtained when $L \times K$ is maximized, which is the number of observations in the trajectory matrix. Now, it is easy to see that $L = L_{\max}$ gives the maximum value of the number of entities of the trajectory matrix. Table 1.3 shows an example of the number of observations that will be incorporated in SSA for several different values of L, when $N = 20, 25$ and 30. Then, the maximum number is highlighted in each case, confirming the theoretical conclusion of this section.

1.4.2 Grouping

A scree graph is a plot of λ_j against j. Alternatives to the scree graph of eigenvalues are to plot $log(\lambda_j)$ and $\sqrt{\lambda_j}$, rather than λ_j, against j.

1 UNIVARIATE SINGULAR SPECTRUM ANALYSIS 23

Table 1.3 Number of observations used in SSA for different L and $N = 20, 25, 30$

N	No. observations when L is equal to															
	L=5	L=6	L=7	L=8	L=9	L=10	L=11	L=12	L=13	L=14	L=15	L=16	L=17	L=18	L=19	L=20
20	80	90	98	104	108	110	110	108	104	98	90	80	68	54	38	20
25	105	120	133	144	153	160	165	168	169	168	165	160	153	144	133	120
30	130	150	168	184	198	210	220	228	234	238	240	240	238	234	228	220

The R function Sing.plt (which is defined below) can be applied for plotting eigenvalues against their number and can easily be altered to plot other alternatives by changing lambda<-svd(X)d to lambda<-log (svd(X))d or using the assignment lambda<-sqrt(svd(X)d) in this function. The plotting procedure is achieved by the following commands:

Program 1.7 Singular values graph R code

```
Sing.plt<-function(Y,L){
lambda<-log(SVD(Y,L)$d)
d<-length(lambda)
win<-1:d
plot.new()
plot.window(xlim=range(win),ylim=range(lambda))
usr=par("usr")
rect(usr[1],usr[3],usr[2],usr[4])
lines(win,lambda,lwd=2)
points(win,lambda,pch=21,bg="gray")
axis(1)
axis(2)
box()
title(xlab="Number")
title(ylab="Log. Singular Values")}
```

Figure 1.6 shows a sample of logarithm of singular values drawn from a random Gaussian model with a mean of zero and a variance of 1, which was produced by the following commands. The Y-axis shows the magnitude of λ:

```
yt<-rnorm(200)
Sing.plt(yt,100)
```

The same can be plotted for the data in Example 1.7 which is the energy series, considering $L = 24$, see Fig. 1.7.

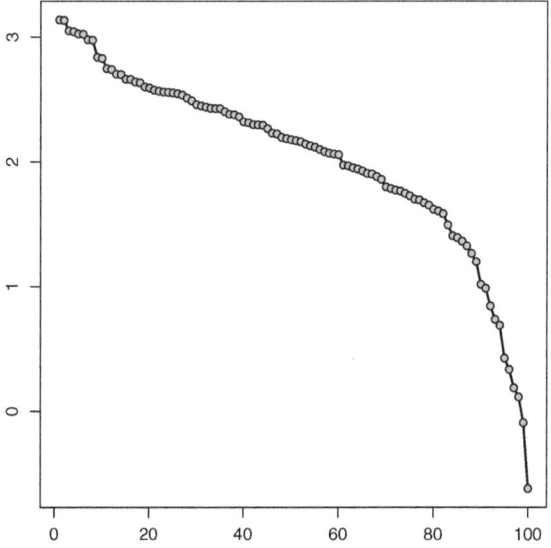

Fig. 1.6 Logarithms of the 200 simulated series eigenvalues

```
Sing.plt(energy,24)
```

The singular values of the two eigentriples of a harmonic series are often very close to each other, and this fact simplifies the visual identification of the harmonic components. An analysis of the pairwise scatterplots of the singular vectors allows one to visually identify those eigentriples that corresponds to the harmonic components of the series. The pure sine and cosine with equal frequencies, amplitudes, and phases create the scatterplot with the points lying on a circle. If $P = 1/w$ is an integer, then these points are the vertices of the regular P-vertex polygon. For the rational frequency $w = m/n < 0.5$, with relatively prime integer m and n, the points are the vertices of the scatterplots of the regular n-vertex polygon.

The following R codes can be applied to draw paired eigenfunctions of the trajectory matrix (U_i versus U_j):

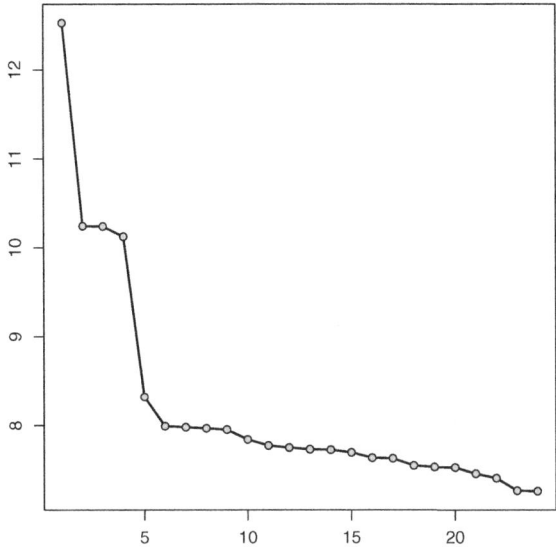

Fig. 1.7 Logarithms of the 24 singular values of the energy series

Program 1.8 Paired eigenfunctions graph R code

```
SSA.Eigen<-function(Y,L){
N<-length(Y)
X<-UniHankel(Y,L)
E<-svd(X)
d<-length(E$d[E$d>1.e-4])
list(eigen.value=E$d,eigen.vector=E$u[,1:d])
}

eigen<-function(Y,L,...){
E<-SSA.Eigen(Y,L)
U<-E$eigen.vector
D<-round(100*E$eigen.value/sum(E$eigen.value),3)
Ev.num<-ncol(U)
windows(width=7,rescale="fit",title="Eigen Functions")

par(mfrow=c(3,3))
n.plot<-1
i<-1
```

```
while(i<Ev.num){
plot(U[,i],U[,(i+1)],type="l",lwd=2,
xlab=substitute(paste(U[s]),list(s=i)),
ylab=substitute(paste(U[s]),list(s=(i+1))))
title(substitute(paste(r ,"(",d1,"%",")","
- ",s,"(",d2,"%",")"),list(r=i,s=(i+1),
d1=D[i],d2=D[(i+1)])),line=1.2,font=2)
n.plot<-n.plot+1
i<-i+1
if(n.plot>9){
windows(width=7,rescale="fit",title="Eigen Functions")
par(mfrow=c(3,3))
n.plot<-1
}}}
```

Using this function, the eigenfunctions for the US energy series can be obtained by the following command:

```
eigen(energy,24)
```

When applying this command, three windows will appear. Here we show the graphs for the first ten paired eigenvectors in Fig. 1.8. Note that, in this example $L = 24$ is considered and the series length is 171, therefore 24 eigenfunctions will be plotted in the graphs.

The methods discussed below are presented to help us achieve a suitable grouping. Although further research is required in establishing some general principles in this area. The following approaches are available for grouping in SSA.

- scree plot of eigenvalues,
- Eigenfunctions and principal components,
- w-correlations among possible groups.

These rules are very much ad hoc rules-of-thumb whose justification, despite some attempts to present them on a more formal basis, is still mainly that they are intuitively plausible and that they work in practice.

1 UNIVARIATE SINGULAR SPECTRUM ANALYSIS   27

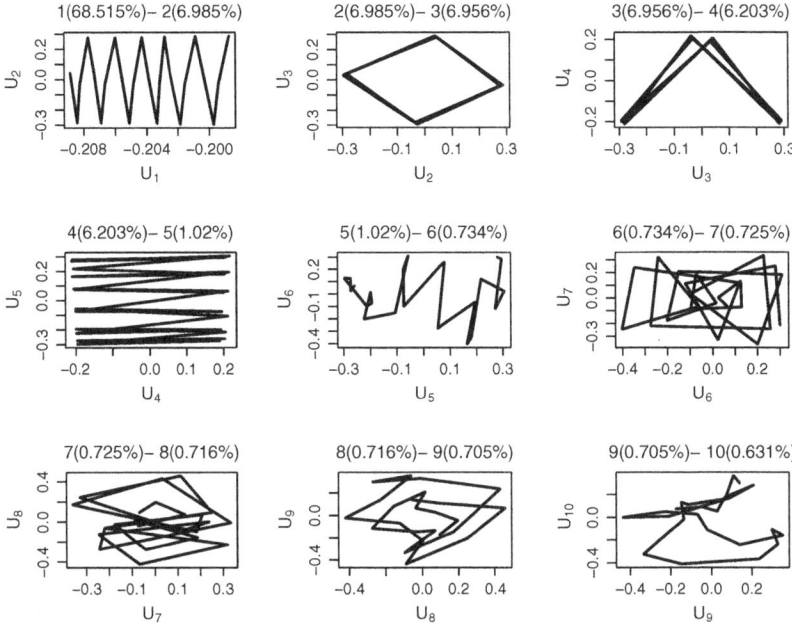

Fig. 1.8 Paired eigenfunctions 1–10 for the energy series

1.5 Forecasting by SSA

The basic requirement to be able to perform SSA forecasting, is that the series satisfies a linear recurrent formula (LRF). The series $Y_N = [y_1, \ldots, y_N]$ satisfies a LRF of order $L - 1$ if:

$$y_t = a_1 y_{t-1} + a_2 y_{t-2} + \cdots + a_{L-1} y_{t-L+1}, \quad t = L+1, \ldots, N \quad (1.10)$$

The series governed by LRFs admit natural recurrent continuation since each term of such a series is a linear combination of several preceding terms. Similar to an Autoregressive model, SSA forecasting is based on the weighting previous observations, with weights obtained based on the eigenvector.

Note that, the Eq. (1.10) does not mean that the series must be linear; but it might be a nonlinear form governed by LRF. For instance, let $y_t = t^2$. Then it can be shown easily that:

$$t^2 = 3(t-1)^2 - 3(t-2)^2 + (t-3)^2, \quad t = 4, 5, \ldots$$

As an other example, assume $y_t = \sin\left(\frac{\pi t}{4}\right)$. Then, trigonometric identities would result in:

$$\sin\left(\frac{\pi}{4}t\right) = \sqrt{2}\sin\left(\frac{\pi(t-1)}{4}\right) - \sin\left(\frac{\pi(t-2)}{4}\right), \quad t = 3, 4, \ldots$$

We consider two main versions of the univariate SSA forecasting algorithm, referred to as Recurrent SSA (RSSA), see Golyandina et al. (2001), and Vector SSA (VSSA), see Golyandina et al. (2001). A practical outline of each of these methods is given in the following two subsections.

We denote by U_j^∇ the vector of the first $L-1$ components of the eigenvector U_j and by π_j the last component of U_j ($j = 1, \ldots, r$). Define the coefficient vector $A \equiv \{a_1, \ldots, a_{L-1}\}$ as follows:

$$A \equiv \frac{1}{1-\upsilon^2}\sum_{j=1}^{r}\pi_j U_j^\nabla,$$

where $\upsilon^2 = \sum_{j=1}^{r}\pi_j^2$.

Accuracy Measure

There are several methods to verify the accuracy of the forecasting model based on the behaviour of the forecast errors. Here, there are two kinds of errors:

- In-sample errors, that is reconstruction errors.
- Out-of-sample errors, that is forecasting errors.

Typically the root mean square error (RMSE) is used as a criterion. If the SSA parameters are minimizing the forecast errors, then the RMSE for both the reconstruction and forecasting would be a minimum. This criterion becomes a method to choose between different values for the SSA parameters. Mahmoudvand et al. (2013) have considered this issue by means of simulation and showed that there are a meaningful differences among the optimal values of window length that are used for reconstruction and forecasting. They showed that optimal value for reconstruction is again close to $L_{\max}$; however, the optimal value depends on the several conditions for the forecasting procedure.

1.5.1 Recurrent Forecasting Method

The RSSA forecasts $(\hat{y}_{N+1},\ldots,\hat{y}_{N+h})$ are obtained by the following recursive formula. Note that here the in-sample data extends to N; therefore, the out-of-sample forecasting at horizon h refers to $N+1,\ldots,N+h$:

$$\hat{y}_i = \begin{cases} \tilde{y}_i, & i = 1,\ldots,N \\ A^T(\hat{y}_{i-L+1},\ldots,\hat{y}_{i-1}) & i = N+1,\ldots,N+h. \end{cases} \quad (1.11)$$

where $\tilde{y}_1,\ldots,\tilde{y}_N$ are the values of the reconstructed series, see Eq. (1.3). The RSSA forecasts are obtained using the following R function:

Program 1.9 Recurrent forecasting SSA R code

```
RSSA.Forecasting<-function(L,groups,h,Y){
N<-length(Y)
L<-min(L,(N-L+1))
X<-UniHankel(Y,L)
U<-matrix(svd(X)$u,L,L)
pi<-array(U[L,groups],dim=length(groups))
V2<-sum(pi^2)
m<-length(groups)
Udelta<-array(U[1:(L-1),groups],dim=c((L-1),m))
A<-pi%*%t(Udelta)/(1-V2)
yhat<-array(0,dim=(N+h))
yhat[1:N]<-SSA.Rec(Y,L,groups)$Approximation
for(l in (N+1):(N+h))
yhat[l]<-A%*%yhat[(l-L+1):(l-1)]
yhat[(N+1):(N+h)]
}
```

Example 1.9 Recurrent forecasting in a real data set

Consider the US energy series again. We assume the cut off point is at $N = 159$ and that we wish to obtain the forecast for the next 12 observations of the energy series using $L = 24$ and the first five eigentriples. These can be obtained using the defined function RSSA.Forecasting. Typical results of this function with $L = 24$ and using the first 5 components are as follows:

```
RSSA.Forecasting(L=24,groups=c(1:5),h=12,Y=energy[1:159])
 [1] 4976.700 6298.907 4102.907 4855.186 4814.090 6110.255
 [7] 3888.206 4638.188 4551.938 5832.808 3602.146 4365.919
```

1.5.2 Vector Forecasting Method

Vector forecasting method is another approach for forecasting in SSA. In general, the vector forecast works more robustly than recurrent especially when faced with outliers or big shocks in the series (Hassani et al. 2014).

Consider the following matrix

$$\Pi = U^{\nabla}U^{\nabla T} + (1 - v^2)AA^T. \qquad (1.12)$$

Now, define the linear operator:

$$\mathscr{P}^{(v)} : \mathfrak{L}_r \mapsto \mathbb{R}^L, \qquad (1.13)$$

where $\mathfrak{L}_r = \text{span}\{U_1, \ldots, U_r\}$ and

$$\mathscr{P}^{(v)}Y = \begin{pmatrix} \Pi Y_\Delta \\ A^T Y_\Delta \end{pmatrix}, \; Y \in \mathfrak{L}_r, \qquad (1.14)$$

where Y_Δ is the vector of the last $L - 1$ elements of Y_N. The vector Z_j is defined as follows:

$$Z_j = \begin{cases} \widetilde{X}_j & \text{for } j = 1, \ldots, K \\ \mathscr{P}^{(v)} Z_{j-1} & \text{for } j = K+1, \ldots, K+h+L-1 \end{cases}, \qquad (1.15)$$

where the $\widetilde{X}_j$s are the reconstructed columns of the trajectory matrix of the ith series after grouping and leaving out noise components. Now, by constructing matrix $Z = [Z_1, \ldots, Z_{K+h+L-1}]$ and performing diagonal averaging, a new series $\hat{y}_1, \ldots, \hat{y}_{N+h+L-1}$ is obtained, where $\hat{y}_{N+1}, \ldots, \hat{y}_{N+h}$ forms the h terms of the VSSA forecast.

In this case the function VSSA.Forecasting is provided for implementing vector forecasting in R as follows:

Program 1.10 Vector forecasting SSA R code

```
VSSA.Forecasting<-function(L,groups,h,Y){
N<-length(Y)
L<-min(L,(N-L+1))
k<-N-L+1
X<-UniHankel(Y,L)
Lambda<-diag(svd(X)$d)
U<-matrix(svd(X)$u,L,L)
V<-matrix(svd(X)$v,k,L)
yhat<-array(U[1:L,groups],dim=c(L,length(groups)))%*%
```

```
Lambda[groups,groups]%*%array(t(V[1:k,groups]),
dim=c(length(groups),k))
pi<-array(U[L,groups],dim=length(groups))
V2<-sum(pi^2)
Udelta<-array(U[1:(L-1),groups],dim=c((L-1),
length(groups)))
A<-pi%*%t(Udelta)/(1-V2)
pai<-Udelta%*%t(Udelta)+(1-V2)*t(A)%*%A
Pv<-rbind(pai,A)
for(l in (k+1):(N+h))
yhat<-cbind(yhat,(Pv%*%yhat[2:L,(l-1)]))
DiagAver(yhat)[(N+1):(N+h)]
}
```

Example 1.10 Vector forecasting in a real data set

Here we forecast the last 12 observations ($h = 12$) of the US energy series using the first 159 observations ($N = 159$) by running the function VSSA.Forecasting, with $L = 24$ and using the first 5 components, as follows:

```
VSSA.Forecasting(L=24,groups=c(1:5),h=12,Y=energy[1:159])
 [1] 4927.785 6249.437 4043.204 4786.854 4749.943 6064.920
 [7] 3840.005 4595.891 4520.599 5826.900 3580.564 4346.145
```

The resulting forecasts from Examples 1.9 and 1.10 are shown below in Fig. 1.9. This figure is produced by the following R code and shows a small difference between VSSA and RSSA as there are no outliers.

```
RSSA<-RSSA.Forecasting(L=24,groups=c(1:5),h=12,Y=energy[1:159])
VSSA<-VSSA.Forecasting(L=24,groups=c(1:5),h=12,Y=energy[1:159])
colr<-c("black","gray","gray50")
ts.plot(as.ts(energy[160:171]),RSSA,VSSA,lty=c(1,1,2),col=colr)
legend("topleft",c("Real","RSSA","VSSA"),lty=c(1,1,2),col=colr)
```

1.5.3 A Theoretical Comparison of RSSA and VSSA

Let us first provide a useful lemma.

Lemma 1.1 *Considering the notations in RSSA and VSSA, the coefficient vector A and projection matrix* Π *satisfy the following equalities:*

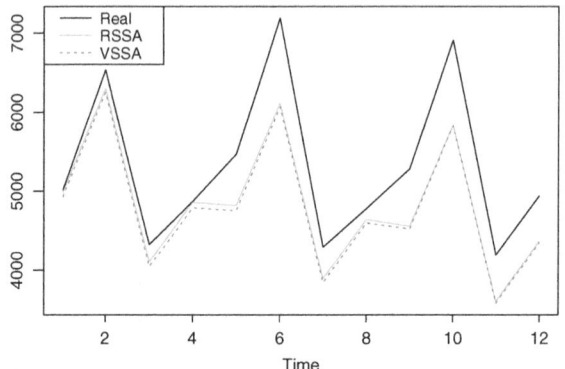

Fig. 1.9 Forecasts from Examples 1.6 and 1.7

$$A = \left(\mathbf{U}^\nabla \mathbf{U}^{\nabla T}\right)^{-} \mathbf{U}^\nabla U_\Delta^T \qquad (1.16)$$

$$\Pi = \mathbf{U}^{\nabla T} \left(\mathbf{U}^\nabla \mathbf{U}^{\nabla T}\right)^{-} \mathbf{U}^\nabla \qquad (1.17)$$

where $\mathbf{U}^\nabla = [U_1^\nabla \ldots U_r^\nabla]$, $U_\Delta = [\pi_1 \ldots \pi_r]$ *and* $\mathbf{Z}^-$ *denotes the generalized inverse of matrix* $\mathbf{Z}$.

According to the above lemma it can be concluded that both RSSA and VSSA use the same projection. Note that in RSSA, we first perform diagonal averaging and then continue the series by LRF to obtain forecasts. However, in VSSA, we first continue the columns by the projection matrix and then use diagonal averaging to obtain forecasts. Therefore, RSSA allows the use of more previous data than VSSA. In RSSA, all entities under the main off-diagonal (the off-diagonal of a matrix running from the upper right entry) are used to obtain forecasts. There are $\frac{L(L-1)}{2}$ entities under the main off-diagonal. In contrast, VSSA uses only the last column which has L observations. This might be a reason why VSSA is more robust than RSSA. In general, the difference between RSSA and VSSA consists in the difference between the last column of the approximated trajectory matrix before and after diagonal averaging. If these are close to each other, then RSSA and VSSA perform equivalently; but if there is a significant difference one should not expect equivalent results.

As an example, run the following R codes to see the discrepancy between the last column of approximated trajectory matrix, before and after diago-

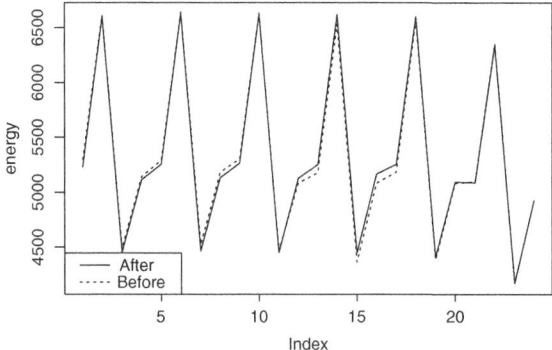

Fig. 1.10 Comparing the last column of the approximated trajectory matrix, before and after diagonal averaging, for the US energy data

nal averaging in Example US energy consumption. The results are demonstrated in Fig. 1.10.

```
after<-SSA.Rec(L=24,Y=energy[1:159],groups=1:5)
after<-after$Approximation[136:159]
before<-Group(L=24,Y=energy[1:159],groups=1:5)[,136]
plot(after,type="l",ylab="energy")
lines(before,type="l",lty=2)
legend("bottomleft",c("After","Before"),lty=1:2)
```

1.6 Automated SSA

In this section we present the automated SSA algorithms for VSSA and RSSA. In each case, there are two relevant R codes. The first code seeks to determine the optimal L and r for a given data set, optimized to minimize a loss function, whilst the second code enables the user to input the optimal L and r in order to generate the forecasts. Note that these codes are applicable to recursive forecasting exercises; that is, where users are interested in obtaining, for example, $h = 1$-step ahead forecasts for a period of one or 2 years recursively.

The automated RSSA and VSSA algorithms are introduced below.

1. Consider a real-valued nonzero time series $Y_N = (y_1, \ldots, y_N)$ of length N.

2. Divide the time series into two parts; say, for example, $\frac{2}{3}$rd of observations for model training and testing, and $\frac{1}{3}$rd for validating the selected model.
3. Use the training data to construct the trajectory matrix $\mathbf{X} = (x_{ij})_{i,j=1}^{L,K} = [X_1, \ldots, X_K]$, where $X_j = (y_j, \ldots, y_{L+j-1})^T$ and $K = N - L + 1$. Initially, we begin with $L = 2$ ($2 \leq L \leq \frac{N}{2}$) and in the process, evaluate all possible values of L for Y_N.
4. Obtain the SVD of $\mathbf{X}$ by calculating $\mathbf{X}\mathbf{X}^T$ for which $\lambda_1, \ldots, \lambda_L$ denotes the eigenvalues in decreasing order ($\lambda_1 \geq \cdots \lambda_L \geq 0$) and by $U_1, \ldots, U_L$ the corresponding eigenvectors. The output of this stage is $\mathbf{X} = \mathbf{X}_1 + \cdots + \mathbf{X}_L$ where $\mathbf{X}_i = \sqrt{\lambda_i} U_i V_i^T$ and $V_i = \mathbf{X}^T U_i / \sqrt{\lambda_i}$.
5. Evaluate all possible combinations of r ($1 \leq r \leq L-1$) singular values (step by step) for the selected L and split the elementary matrices $\mathbf{X}_i$ ($i = 1, \ldots, L$) into several groups and sum the matrices within each group.
6. Perform diagonal averaging to transform the matrix with the selected r singular values into a Hankel matrix, which can then be converted into a time series (the steps up to this stage filter the noisy series). The output is a filtered series that can be used for forecasting.
7. Depending on the forecasting approach one wishes to use, select either the RSSA approach or the VSSA approach, which are explained in Sects. 1.4.1 and 1.4.2, respectively.
8. Define a loss function $\mathscr{L}$.
9. When forecasting a series Y_N h-step ahead, the forecast error is minimized by setting $\mathscr{L}(X_{K+h} - \hat{X}_{K+h})$ to be minimum where the vector $\hat{X}_{K+h}$ contains the h-step ahead forecasts obtained using the RSSA/VSSA forecasting algorithm.
10. Find the combination of L and r which minimized $\mathscr{L}$ and thus represents the optimal VSSA choices.
11. Finally, use the optimal L to decompose the series comprising of the training set, and then select r singular values for reconstructing the less noisy time series, and use this newly reconstructed series for forecasting the remaining $\frac{1}{3}$rd observations.

Here, we begin by presenting the a automated RSSA code for finding the optimal RSSA parameters for a given data set. The output from the R code presented in Program (1.13), which must be executed after running Programs (1.1)–(1.4) and Programs (1.9)–(1.12), is a list which will show the r and L minimizing RMSE and RMSE attainable for the best r

combined with each L. The L and r which corresponds to the minimum RMSE will provide the optimal L and r for obtaining forecasts for a given data set.

In this function, users need to define L.vect which denotes L and ($2 \leq L \leq N/2$) where N is the length of the training set in your time series. Then use the h to select the forecasting horizon of interest, and finally define mse.size which is the number of forecasts (i.e. the test set) for which you wish to determine the optimal SSA parameters. Other loss functions can be used instead of RMSE and in that case it is enough to change the R code in Program (1.11) in such a way that it computes the corresponding loss function and then use the new loss function in lines 4 and 6 with Programs (1.12) and (1.13), respectively.

Program 1.11 RMSE R code for SSA forecasting

```
RMSE<-function(Y,L,r,mse.size,h){
N<-length(Y)
forecasts<-NULL
train.size<-N-mse.size-h+1
test<-Y[(train.size+h):(train.size+mse.size+h-1)]
for(i in 1:mse.size){
train<-Y[1:(train.size+i-1)]
forecasts<-c(forecasts,RSSA.Forecasting(L,1:r,h,train)[h])
}
sqrt(mean((test-forecasts)^2))
}
```

Program 1.12 R code for optimal r

```
Opt.r<-function(Y,L,mse.size,h){
opt.rmse<-array(0,dim=(L-1))
for(r in 1:(L-1))
opt.rmse[r]<-RMSE(Y,L,r,mse.size,h)
which.min(opt.rmse)
}
```

Program 1.13 R code for optimal choices

```
Opt.choices<-function(Y,L.vec,h,mse.size){
    wid<-length(L.vec)
    rmse<-array(0,dim=wid)
for(i in 1:wid){
```

```
r0<-Opt.r(Y,L.vec[i],mse.size,h)
rmse[i]<-RMSE(Y,L.vec[i],r0,mse.size,h)
}
window<-which.min(rmse)
r.opt<-Opt.r(Y,L.vec[window],mse.size,h)
L.opt<-L.vec[window]
list(optimal.r=r.opt,optimal.L=L.opt,root.mean.square.error=min(rmse))
}
```

> **Example 1.11 Automated forecasting in a real data set**

Here we find the optimal choices for 12-step ahead forecasting of the US energy series using the first 159 observations by running the functions Opt.choices() *as follows:*

```
Opt.choices(Y=energy[1:159],L.vec=2:24,mse.size=10,h=12)
$optimal.r
[1] 5

$optimal.L
[1] 24

$root.mean.square.error
[1] 350.0272
```

The optimal pair (L, r) is $(24, 5)$ which produces an RMSE equal to 350.03. Let us explain how RMSE is computed in the above example. The length of series is 159, mse.size equals 10 and h=12. We begin with the first 138 observations and find the forecast for observation 150, we then consider the first 139 observations and obtain the forecast for observation 151 and continue until the forecast of the last observation, 159. In this way, we obtain ten 12-step ahead forecasts. Comparing these forecasts with test set, observations 150–159, we obtain the RMSE.

1.6.1 Sensitivity Analysis

This section investigates the sensitivity of results of SSA forecasts to the selection of r and L. We can use a three-dimensional graph showing RMSE on the vertical axis and L and r on the horizontal axis; but, 3D graphs pose some challenges to data analysis. For example, the representation of points in 3D tends to appear more cluttered, making it more difficult to interpret the data accurately. Note that Program 1.14, which requires Program 1.11, can be used to produce a 3D graph for a range of L and r. It is worth mentioning that, there are several useful packages and R functions to produce 3D graphs in R, however, we used the R function `persp()` which does not require any new packages to be installed for R.

Program 1.14 Sensitivity analysis R code

```
all.rmse<-function(Y,L,mse.size,h,max.r){
rmse<-array(0,dim=max.r)
for(r in 1:max.r)
rmse[r]<-RMSE(Y,L,r,mse.size,h)
rmse
}

Sensitivity<-function(Y,L.vec,h,mse.size,max.r,theta,phi){
    wid<-length(L.vec)
    r.val<-1:max.r
    rmse<-array(0,dim=c(wid,max.r))
for(i in 1:wid){
rmse[i,]<-all.rmse(Y,L.vec[i],mse.size,h,max.r)
}
persp(y=r.val,x=L.vec,z=rmse,theta=theta,phi=phi,shade=.10)
}
```

Example 1.12 Sensitivity of forecasting to r and L in US energy data

We now use Program 1.14 for the US energy data with the following R code resulting in Fig. 1.11:

```
Sensitivity(Y=energy[1:159],L=15:24,mse.size=10,h=12,max.r=10
+,theta=80,phi=15)
```

we obtain a 3D graph as shown in Fig. 1.11. This implies again that $L = 24$ and $r = 5$ provide minimum RMSE.

Figure 1.11 is very informative as enables us to search for optimal values of L and r simultaneously. It also aids a search for local minimum for a particular value of either L or r. For instance, for a fixed value of L, one is able to find a proper value of r such that RMSE is minimum. Similar information can be obtained for a fixed value of r and different values of L.

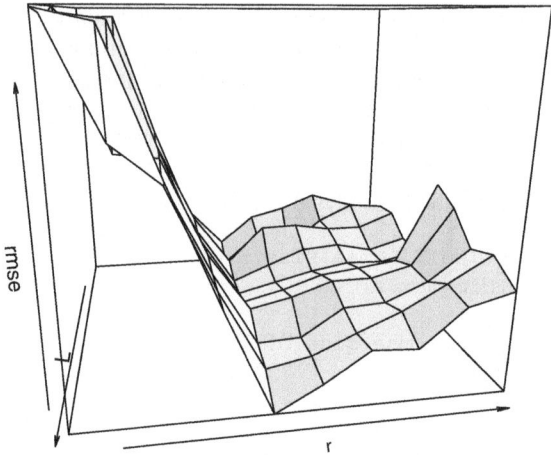

Fig. 1.11 Sensitivity analysis of RMSE of forecasts in US energy data

1.7 PREDICTION INTERVAL FOR SSA

Prediction intervals can be very useful in assessing the quality of the forecasts. Unlike the SSA forecasts themselves (their construction does not formally require any preliminary information about the time series), current prediction bounds need some assumptions to be imposed on the series and the residual component, which we associate with noise. Let $Y_N = S_N + E_N$, where $S_N = (s_1, \ldots, s_N)'$ and $E_N = (e_1, \ldots, e_N)'$ be the signal and noise components of the series Y_N, respectively. In this case, see Golyandina et al. (2001), considered two alternatives for constructing the confidence bounds for SSA forecasts:

- Empirical prediction intervals, where the intervals are obtained for the entire series, i.e. both the signal and noise components:

$$y_{N+h} \in \left(\hat{y}_{N+h} + c_{\alpha/2},\ \hat{y}_{N+h} + c_{1-\alpha/2} \right). \tag{1.18}$$

- Bootstrap prediction intervals, where the intervals are obtained for the signal component of the series:

$$s_{N+h} \in \left(\hat{s}_{N+h,\alpha/2},\ \hat{s}_{N+h,1-\alpha/2} \right). \tag{1.19}$$

Here, $c_{\alpha/2}, c_{1-\alpha/2}, \hat{s}_{N+h,\alpha/2}$ and $\hat{s}_{N+h,1-\alpha/2}$ are quantities that depend on the confidence level $1 - \alpha$.

The first prediction interval method Eq. (1.18) exploits information about the forecast errors (residual series) obtained by processing the series. Under the assumption that the series of residuals is stationary and ergodic, the quantiles of the related marginal distribution can be estimated and the confidence bounds can be built. This variant of obtaining the prediction intervals is called the *empirical method*.

The alternative prediction interval from Eq. (1.19) requires additional information about the model governing the time series that includes the noise components, E_T, to accomplish a bootstrap simulation of the series $Y_T = S_T + E_T$. Bootstrap prediction intervals are calculated for the reconstructed time series and its forecast, assuming that the initial time series is equal to the sum of a signal and a normal white noise, and that the reconstructed time series approximately coincides with the signal.

In the remainder of this section we, give a brief description of the second method as it is more realistic to use in practice. Further details can be found in Golyandina et al. (2001).

Bootstrap Based Prediction Interval for SSA Forecasting

To obtain the bootstrap prediction interval in Eq. (1.19) for the h-step ahead forecast, the first step is to obtain the SSA decomposition $Y_N = \widetilde{S}_N + \widetilde{E}_N$, where $\widetilde{S}_N$, the reconstructed series, approximates the signal S_N of the time series and $\widetilde{E}_N$ is the residual series. Assuming that we have a specific model for the residuals $\widetilde{E}_N$, we simulate p independent copies $\widetilde{E}_{N,i}$, $i = 1, \ldots, p$, of the residual series E_N. Adding each of these residual series to the signal series $\widetilde{S}_N$, we get p series $Y_{N,i} = \widetilde{S}_N + \widetilde{E}_{N,i}$. Applying the Recurrent (or Vector) SSA forecasting algorithm, keeping unchanged the window length L and the number r of eigenvalues/eigenvectors used for reconstruction, to the series $Y_{N,i}, i = 1, \ldots, p$, we can obtain p forecasting results h-step ahead $\hat{y}_{N+h,i}$.

The empirical $\alpha/2$ and $1 - \alpha/2$ quantiles of the p h-step ahead forecasts $\hat{y}_{N+h,1}, \ldots \hat{y}_{N+h,p}$, correspond to the bounds of the bootstrap prediction interval with confidence level $1-\alpha$. Program (1.15) can be applied to obtain the bootstrap-based prediction interval for Recurrent SSA forecasts.

Program 1.15 Bootstrap based RSSA prediction interval

```
Pred.int<-function(L,groups,h,Y,p,alpha){
N<-length(Y)
Recons<-SSA.Rec(Y,L,groups)
S.hat<-Recons$Approximation
E.hat<-Recons$Residual
```

```
L.pi<-U.pi<-0
new<-array(dim=c(h,p))
for(i in 1:p){
Y.hat<-S.hat+sample(E.hat,N,replace=TRUE)
new[,i]<-RSSA.Forecasting(L,groups,h,Y.hat)
}
L.pi<-apply(new,1,quantile,probs=alpha)
U.pi<-apply(new,1,quantile,probs=1-alpha)
data.frame(L.pi,U.pi)
}
```

As an example, using Program (1.15) we obtained the 95% prediction interval for forecasting the 6 last observations of the US energy series as below:

```
Pred.int(24,1:15,6,energy[1:167],1000,.05)
    L.pi      U.pi
1 5098.137  5696.939
2 6285.734  7070.748
3 3475.925  4171.856
4 4545.088  5297.123
5 5034.574  5987.344
6 5987.567  6880.463
```

Let us now consider the application of all the above-mentioned procedures for analysing two real data sets.

1.8 Two Real Data Analysis by SSA

1.8.1 UK Gas Consumption

This time series is of quarterly UK gas consumption from 1960Q1 to 1986Q4, in millions of therms. The length of time series is $N = 108$. In addition, the package of data in R includes this time series and we can see the observations by writing the R code UKgas in the command line. The following can then be performed SSA provided that the R codes with Programs (1.1)–(1.13) are defined in R. In the first stage, apply the following R code to produce a plot of the UK gas series as shown in the left plot of Fig. 1.12.

```
plot.ts(UKgas)
```

1 UNIVARIATE SINGULAR SPECTRUM ANALYSIS   41

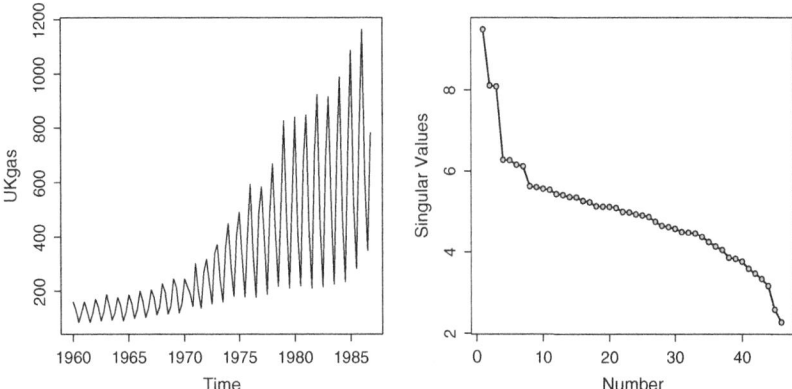

Fig. 1.12 Quarterly UK gas consumption time series over the period 1960Q1–1986Q4

Then, divide the time series into two parts: for instance the first 92 observations for training and model building and the last 16 observations for testing and validation. Use the following code to see how the singular values of the trajectory matrix behave. This code produces the right plot in Fig. 1.12. This plot shows that the first 7 components might be sufficient to reconstruct the main time series.

```
Sing.plt(UKgas[1:92],46)
```

But, let us to look at the eigenfunctions with the following R code.

```
eigen(UKgas[1:92],46)
```

Note that this code produces 5 graphic windows, each with 9 plots, but we only demonstrate the first graphic windows including the first nine eigenfunctions, see Fig. 1.13. The first 7 components have useful information and the remaining components can be considered as noise.

Another useful tool to evaluate the appropriateness of the grouping is the w-correlation measure that can be calculated for the UK gas series by the following R code:

```
m<-W.corr(Yt=UKgas[1:92],L=46,groups=seq(1:45))
Plt.Img(m)
title(main="w-correlation")
```

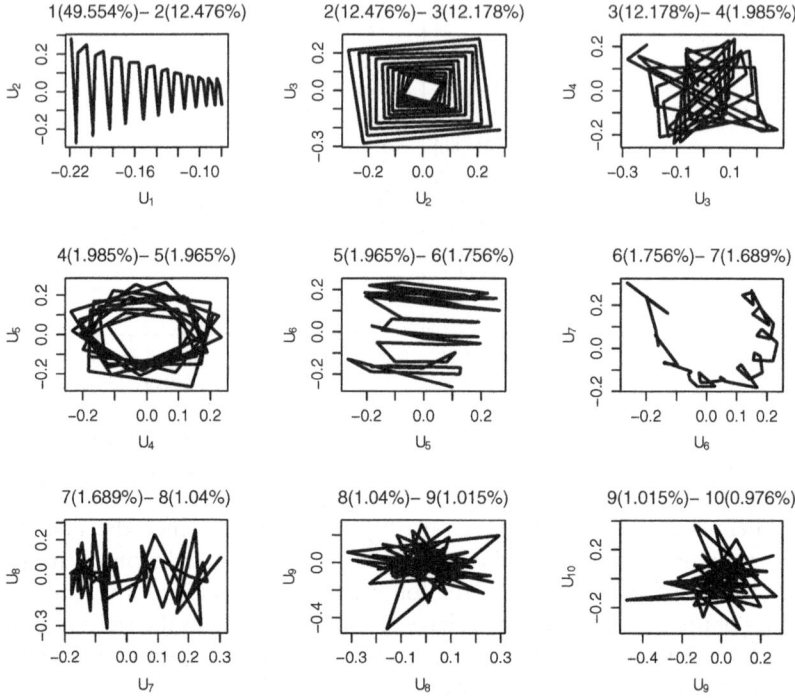

Fig. 1.13 First nine Eigenfunctions for UK gas time series with $L = 46$

The results of this code, are shown in Fig. 1.14. This figure confirms the previous conclusion, implying that the seven first components are signal extractors.

The above analysis was based on $L = 46$, where as we now use the automated RSSA code to find the optimal choices for forecasting, say, 4-step ahead. Using the following R code results in the pair $(L = 39, r = 11)$ for obtaining the 4-step ahead forecasts.

```
Opt.choices(Y=UKgas[1:92],L.vec=2:41,mse.size=5,h=4)
$optimal.r
[1] 11

$optimal.L
[1] 39

$root.mean.square.error
[1] 22.97463
```

1 UNIVARIATE SINGULAR SPECTRUM ANALYSIS   43

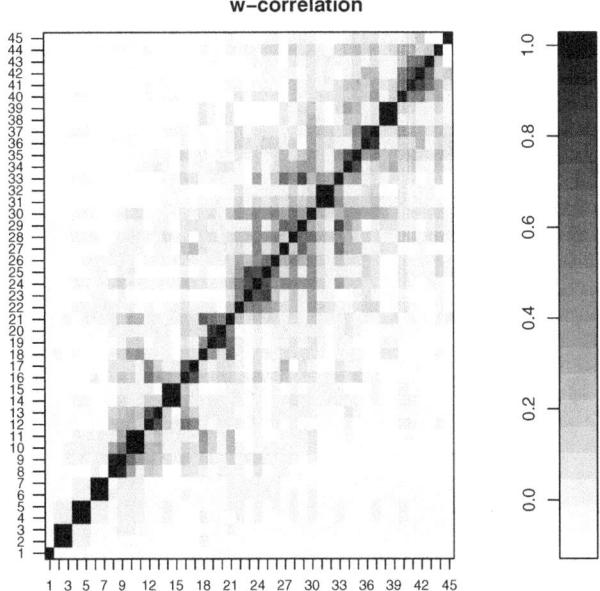

Fig. 1.14 W-correlations among pair components for UK gas time series with $L = 46$

We now compare the forecasting results from the two different pairs of SSA choices: (i): $L = 46, r = 7$ and (ii) $L = 39, r = 11$. In order to do that, we have used the following R code, which will produce the plot in Fig. 1.15.

```
graphbase<-automated<-0
for(t in 1:13){
Yt<-UKgas[1:(92+t-1)]
 graphbase[t]<-RSSA.Forecasting(L=46,1:7,h=4,Yt)[4]
 automated[t]<-RSSA.Forecasting(L=39,1:11,h=4,Yt)[4]
}
rang<-range(UKgas[96:108],graphbase,automated)
plot(96:108,UKgas[96:108],ylim=rang,type="o")
lines(96:108,automated,type="o",pch=15,lty=2)
lines(96:108,graphbase,type="o",pch=17,lty=3)
```

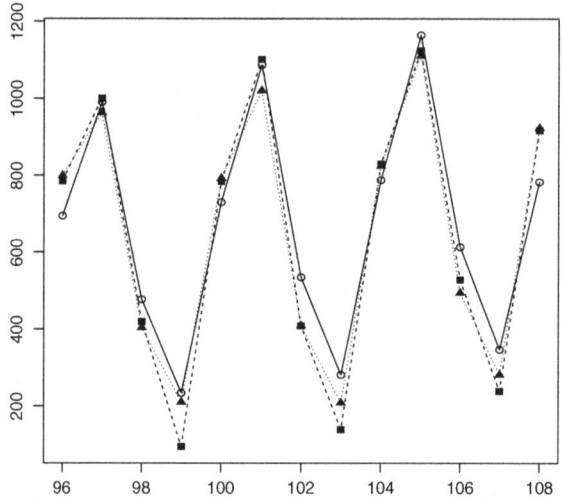

Fig. 1.15 Comparison of forecasts by paris (i) ($L = 46, r = 7$) and (ii) ($L = 39, r = 11$) for UK gas consumption series; solid line with circle points show the original time series, dashed lines with triangle and square symbols show the forecasts by (i) and (ii), respectively

1.8.2 The Real Yield on UK Government Security

This time series is the monthly UK government security yield from Jan. 1985–Dec. 2015. The length of time series is 372. We have entered this data by using the R function scan() and assigning name UKYield to it. We can now undertake the SSA, provided that the R codes with Programs (1.1)–(1.13) are defined in R. In the first stage, apply the following R code to produce a plot of the UKYield series as visible via the left plot in Fig. 1.16.

```
UKYield<-ts(UKYield,start=c(1985,1),freq=12)
plot(UKYield,ylab="UK Government Security Yield")
```

Then, divide the time series into two parts: for instance the first 350 observations for training and model building and the last 22 observations for testing and validation. Use the following code to see how the singular values of the trajectory matrix behave. This code produces the right plot in Fig. 1.16. This plot shows that the first 13 components might be sufficient to reconstruct the main time series.

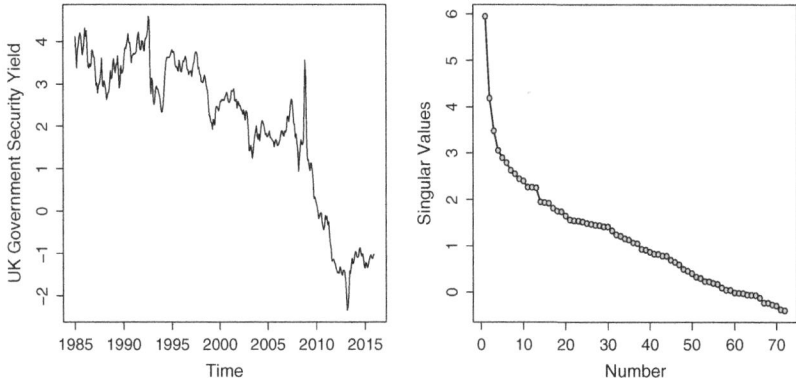

Fig. 1.16 Monthly UK government security yield time series over the period Jan. 1985–Dec. 2015

```
Sing.plt(UKYield[1:350],72)
```

Additionally, we consider at the eigenfunctions with the following R code.

```
eigen(UKYield[1:350],72)
```

Note that this code produces 8 graphic windows, each with 9 plots, but we only demonstrate the first graphic window including the first nine eigenfunctions, see Fig. 1.17.

As in the previous example, we check the suitability of the grouping with the w-correlation measure using the following R code:

```
m<-W.corr(Yt=UKYield[1:350],L=72,groups=seq(1:51))
Plt.Img(m)
title(main="w-correlation")
```

The results of this code are shown in Fig. 1.18. This figure confirms the previous conclusion implying that the seven first components are the signal extractors.

The above analysis was based on $L = 72$ and as in previous example, we now use the automated RSSA code to find the optimal choices for forecasting, for instance, 22-step ahead. Using the following R code results in the pair $(L = 96, r = 3)$ for obtaining the 22-step ahead forecasts.

```
Opt.choices(Y=UKYield[1:350],L.vec=seq(24,156,12),mse.size=5,h=22)
$optimal.r
```

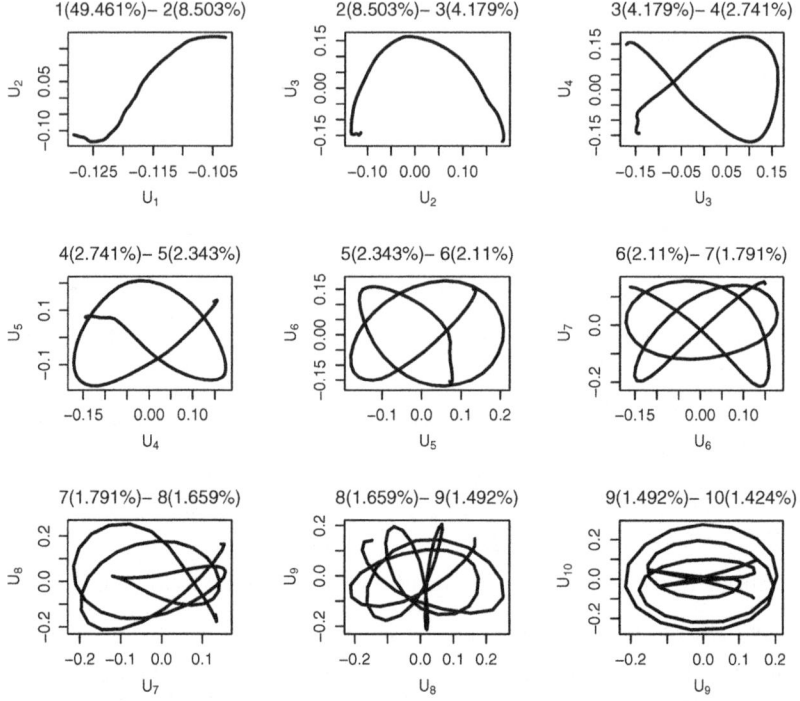

Fig. 1.17 First nine Eigenfunctions for UK government security yield time series with $L = 72$

```
[1] 3

$optimal.L
[1] 96

$root.mean.square.error
[1] 0.132547
```

Finally, we compare the forecasting results for the two different pairs of SSA choices: (i): $L = 72, r = 13$ and (ii) $L = 96, r = 3$. The related R code is as follows which will produce the plot in Fig. 1.19.

```
graphbase<-automated<-0
for(t in 1:11){
```

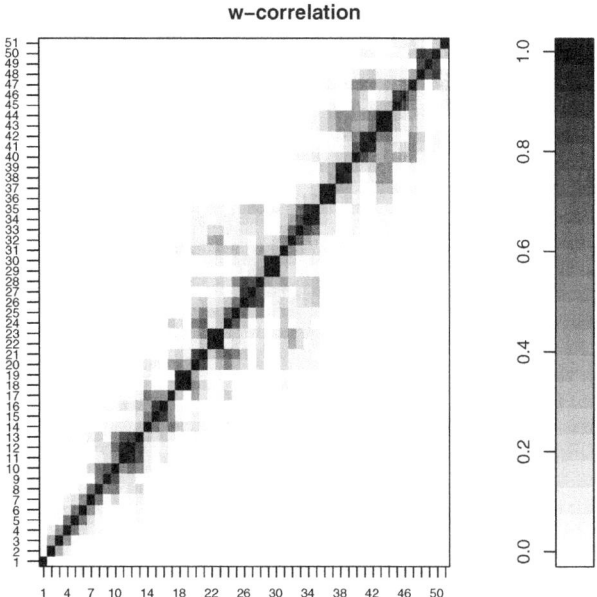

Fig. 1.18 W-correlations among pair components for UK government security yield time series with $L = 72$

```
Yt<-UKYield[1:(350+t-1)]
graphbase[t]<-RSSA.Forecasting(L=72,1:13,h=12,Yt)[4]
automated[t]<-RSSA.Forecasting(L=96,1:3,h=12,Yt)[4]
}
rang<-range(UKYield[362:372],graphbase,automated)
plot(362:372,UKYield[362:372],ylim=rang,type="o")
lines(362:372,automated,type="o",pch=15,lty=2)
lines(362:372,graphbase,type="o",pch=17,lty=3)
legend("topleft",c("real","automated","user"),pch=c(1,15,17))
```

As can be seen in Fig. 1.19 the automated approach produces more accurate forecasting results compared with those obtained by user selection.

1.9 Conclusion

The R code related to the SSA concept and its related fundamental details were discussed in this chapter including an overview of the main mathemat-

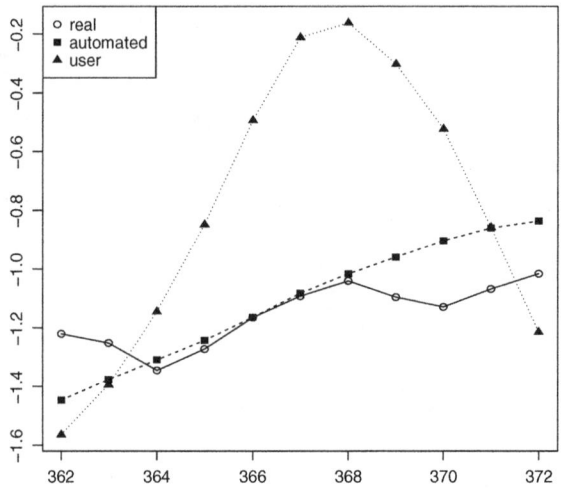

Fig. 1.19 Comparison of forecasts by user ($L = 72, r = 13$) and automated choices ($L = 96, r = 3$) for UK government security yield series

ical applications. The concepts of univariate SSA discussed in this chapter and the output of related R cods, were also illustrated with various real and simulated series.

The second chapter of this book is devoted to multivariate version of SSA (MSSA) which considers four different versions of SSA and the associated R codes for MSSA, with many real and simulated examples are also provided.

CHAPTER 2

Multivariate Singular Spectrum Analysis

Abstract When multiple time series are observed, we are usually interested in the internal structure of each, and at the same time their joint structure, or the dependency among series. Accordingly, the second chapter of this book is dedicated to this vital concept. In this chapter, the basic univariate Singular Spectrum Analysis (SSA) is extended in a fairly obvious way to the multivariate case and transition from univariate SSA to multivariate with emphasis on intuition and applications are deliberated. The related multivariate SSA codes along with several practical examples are also presented.

Keywords Multivariate time series · Multivariate SSA · Block Hankel Matrix · Interdependency

2.1 Introduction

The Singular Spectrum Analysis (SSA) technique can be applied to a single series or jointly to several series and, in the latter case, it is referred to as MSSA. As in the case of parametric modelling, two or more series may be related, which in the context of MSSA has a correspondence in terms of matched components of the several series. There are numerous examples of successful applications of MSSA (e.g. see Patterson et al. (2011) and references therein). Generally speaking, the main applications of MSSA are filtering and forecasting times series. This chapter provides an outline of MSSA together with the R programs necessary to apply this method.

2.2 Filtering by MSSA

The univariate SSA filtering algorithm is based on one Hankel matrix; however, for the multivariate case the trajectory matrix, which is the main tool for SSA, can be defined in different ways.

The discrepancy between the two main approaches in MSSA is mainly due to organization of the single trajectory matrix $\mathbf{X}$ of each series into a block trajectory matrix in the multivariate case. For example, two trajectory matrices can be organized either in vertical or horizontal forms. In this book, multivariate SSA with the vertical form is called VMSSA and with the horizontal form is called HMSSA. Considering these approaches, there are two different filtering methods as follows:

$$\text{MSSA filtering approach} = \begin{cases} \text{HMSSA Horizontal approach} \\ \text{VMSSA Vertical approach} \end{cases}$$

In Sects. 2.2.1 and 2.2.2 we consider the situations where there are M time series with different series length N_i; $Y_{N_i}^{(i)} = (y_1^{(i)}, \ldots, y_{N_i}^{(i)})$ ($i = 1, \ldots, M$).

As with univariate SSA, filtering by MSSA consists of two complementary stages: decomposition and reconstruction. In the first stage, the series is decomposed and in the second stage, the noise-free series is reconstructed. In what follows, we will present details with the R codes in order to use HMSSA and VMSSA for filtering multivariate time series.

2.2.1 MSSA: Horizontal Form (HMSSA)

Stage I. Decomposition

Step 1: Embedding. As noted in Chapter 1, Sect. 1.2, the trajectory matrix $\mathbf{X}^{(i)}$ is a Hankel matrix and the embedding procedure for each series separately provides M different $L_i \times K_i$ trajectory matrices $\mathbf{X}^{(i)}$, where L_i ($2 \leq L_i \leq N_i - 1$) is the window length for each series with length N_i and $K_i = N_i - L_i + 1$ ($i = 1, \ldots, M$). The result of this step is the block Hankel trajectory matrix, $\mathbf{X}_H$, given by:

$$\mathbf{X}_H = [\mathbf{X}^{(1)} : \mathbf{X}^{(2)} : \ldots : \mathbf{X}^{(M)}]. \tag{2.1}$$

Thus, $\mathbf{X}_H$ indicates that the output of the first step is a block Hankel trajectory matrix formed in *horizontal* sequence. To form a new block Hankel matrix in a horizontal form, we need to have $L_1 = \ldots, L_M = L$. Accordingly, this version of MSSA enables us to have various K_i and different series length N_i, however, it requires equal L_i for all series. The following R function constructs the trajectory matrix in HMSSA:

Program 2.1 HMSSA Hankel matrix construction R code

```
HMSSA.Hankel<-function(Y,L){
X<-NULL
if(is.list(Y)){
M<-length(Y)
for(i in 1:M){
ki<-length(Y[[i]])-L+1
Xi<-outer((1:L),(1:ki),function(x,y) Y[[i]][(x+y-1)])
X<-cbind(X,Xi)}
}
else{
Y<-as.matrix(Y)
M<-ncol(Y)
for(i in 1:M){
ki<-length(Y[,i])-L+1
Xi<-outer((1:L),(1:ki),function(x,y) Y[(x+y-1),i])
X<-cbind(X,Xi)}
}
X
}
```

In this function, Y is the input argument containing the time series data for all the component series and can be defined as list() for the unequal series or matrix(), when the series lengths are the same. The next lines show two examples of running this code.

> **Example 2.1: Constructing multivariate trajectory matrix when series are of an unequal length**

In this example, we consider a bivariate time series with 6 observations for the first series and 5 observations for the second series:

```
Y<-list(c(1:6),c(-1,0,3,5,7))
HMSSA.Hankel(Y,3)
     [,1] [,2] [,3] [,4] [,5] [,6] [,7]
[1,]   1    2    3    4   -1    0    3
[2,]   2    3    4    5    0    3    5
[3,]   3    4    5    6    3    5    7
```

> **Example 2.2: Constructing multivariate trajectory matrix when series are of an equal length**

Here we consider a bivariate time series with 6 observations for both series:

```
Y<-matrix(c(1:6,-1,0,3,5,7,2),6,2) # nrow=6, ncol=2
HMSSA.Hankel(Y,3)
     [,1] [,2] [,3] [,4] [,5] [,6] [,7] [,8]
[1,]   1    2    3    4   -1    0    3    5
[2,]   2    3    4    5    0    3    5    7
[3,]   3    4    5    6    3    5    7    2
```

Step 2: SVD. In this step, the SVD of $\mathbf{X}_H$ is performed. Denote by $\lambda_{H_1}, \ldots, \lambda_{H_L}$ the eigenvalues of $\mathbf{X}_H \mathbf{X}_H^T$, arranged in decreasing order $(\lambda_{H_1} \geq \cdots \lambda_{H_L} \geq 0)$ and by $U_{H_1}, \ldots, U_{H_L}$, the corresponding eigenvectors. The SVD of $\mathbf{X}_H$ can be written as $\mathbf{X}_H = \mathbf{X}_{H_1} + \cdots + \mathbf{X}_{H_L}$, where

$$\mathbf{X}_{H_i} = \sqrt{\lambda_i} U_{H_i} V_{H_i}^T, \qquad (2.2)$$

and

$$V_{H_i} = \mathbf{X}_H^T U_{H_i} / \sqrt{\lambda}_{H_i} \quad (\mathbf{X}_{H_i} = 0 \ if \ \lambda_{H_i} = 0). \qquad (2.3)$$

Hence, the structure of the matrix $\mathbf{X}_H \mathbf{X}_H^T$ is as follows:

$$\mathbf{X}_H \mathbf{X}_H^T = \mathbf{X}^{(1)} \mathbf{X}^{(1)^T} + \cdots + \mathbf{X}^{(M)} \mathbf{X}^{(M)^T}. \quad (2.4)$$

From the structure of the matrix $\mathbf{X}_H \mathbf{X}_H^T$ in HMSSA, we see that there are no cross products between Hankel matrices $\mathbf{X}^{(i)}$ and $\mathbf{X}^{(j)}$.

Note also that performing the SVD of $\mathbf{X}_H$ in HMSSA yields L eigenvalues as in univariate SSA, whilst we have $L_{sum} = \sum_{i=1}^{M} L_i$ eigenvalues in VMSSA, see Sect. 2.2.2.

Using the R function HMSSA.Hankel(), the following function can be applied to compute the SVD of the trajectory matrix in HMSSA:

Program 2.2 SVD of HMSSA R code

```
H.SVD<-function(Y,L){
X<-HMSSA.Hankel(Y,L)
svd(X)
}
```

The outputs of H.SVD() are the eigenvalues, eigenvectors and principal components of the trajectory matrix. For the same data in Example 2.2, the SVD can be obtained as below:

```
H.SVD(Y,3)
$d
[1] 18.415453   5.791681   1.525623

$u
              [,1]        [,2]        [,3]
[1,]   -0.3961926  -0.5604541  -0.7272706
[2,]   -0.6191664  -0.4218045   0.6623548
[3,]   -0.6779855   0.7127216  -0.1798985

$v
              [,1]         [,2]         [,3]
[1,]   -0.19920672   0.12675107   0.03784906
[2,]   -0.29115909   0.08021243  -0.12261934
[3,]   -0.38311147   0.03367379  -0.28308773
[4,]   -0.47506384  -0.01286485  -0.44355613
[5,]   -0.08893422   0.46594744   0.12294973
[6,]   -0.28494694   0.39680960   0.71287044
[7,]   -0.49036582   0.20696351  -0.08477028
```

[8,] -0.41655772 -0.74753056   0.41971937

The eigentriples could also be obtained by using SSA for each series separately, with the following results:

```
H.SVD(Y[,1],3)
$d
[1]  1.301119e+01  8.419251e-01  1.871627e-16

$u
             [,1]         [,2]         [,3]
[1,]   -0.4176729  -0.8117159   0.4082483
[2,]   -0.5647271  -0.1200692  -0.8164966
[3,]   -0.7117813   0.5715774   0.4082483

$v
             [,1]         [,2]         [,3]
[1,]   -0.2830233   0.7873359  -0.5475034
[2,]   -0.4132328   0.3594977   0.7415524
[3,]   -0.5434424  -0.0683404   0.1594054
[4,]   -0.6736519  -0.4961785  -0.3534544
H.SVD(Y[,2],3)
$d
[1]  13.092830   5.700905   1.038023

$u
             [,1]         [,2]         [,3]
[1,]    0.3773008   0.5548258   0.7414934
[2,]    0.6734605   0.3852138  -0.6309210
[3,]    0.6356847  -0.7374135   0.2283118

$v
             [,1]         [,2]          [,3]
[1,]    0.1168390  -0.4853732  -0.05448616
[2,]    0.3970727  -0.4440394  -0.72368711
[3,]    0.6835037  -0.2756314   0.64358734
[4,]    0.6012525   0.7009060  -0.24311224
```

The comparison between eigentriples of HMSSA and SSA indicates that the results of univariate and multivariate SSA differ, which is, in fact, the effect of the dependency among the time series.

Stage II. Reconstruction

Step 1: Grouping. This step corresponds to splitting the matrices $\mathbf{X}_{H_1}, \ldots, \mathbf{X}_{H_L}$ into several disjoint groups and summing the matrices within each group. The split of the set of indices $\{1, \ldots, L\}$ into disjoint subsets $I_1, \ldots, I_m$ corresponds to the representation $\mathbf{X}_H = \mathbf{X}_{I_1} + \cdots + \mathbf{X}_{I_m}$. As in univariate SSA, the procedure of choosing the sets $I_1, \ldots, I_m$ is called grouping. For a given group I, the contribution of the component $\mathbf{X}_{H_I}$ is measured by the share of the corresponding eigenvalues: $\sum_{i \in I} \lambda_{H_i} / \sum_{i=1}^{d_H} \lambda_{H_i}$, where d_H is the rank of $\mathbf{X}_H$ and $I \subset \{1, \ldots, L\}$. As a simple case, where there are only signal and noise components, two groups of indices are used, $I_1 = \{1, \ldots, r\}$ and $I_2 = \{r+1, \ldots, L\}$ with $r < L$, where the group $I = I_1$ is associated with the signal component and the group I_2 with the noise.

Step 2: Diagonal averaging or Hankelization. The purpose of diagonal averaging is to transform the reconstructed matrix $\mathbf{X}_{I_j}$ to the form of a Hankel matrix, which can be subsequently converted to a time series. Let $\widetilde{\mathbf{X}}^{(i)}$ be the approximation of $\mathbf{X}^{(i)}$ obtained from the diagonal averaging step. If $\widetilde{x}_{mn}^{(i)}$ stands for an element of a matrix $\widetilde{\mathbf{X}}^{(i)}$, then the jth term of the reconstructed series $\widetilde{Y}_{N_i}^{(i)} = (\widetilde{y}_1^{(i)}, \ldots, \widetilde{y}_j^{(i)}, \ldots, \widetilde{y}_{N_i}^{(i)})$ is achieved by arithmetically averaging $\widetilde{x}_{mn}^{(i)}$ over all (m, n) such that $m + n - 1 = j$. Completing this algorithm, we obtain a filtered time series, which then will be used for forecasting. In order to write R codes to perform the grouping step in HMSSA, we can use the following code:

Program 2.3 Grouping of HMSSA R code

```
H.Group<-function(Y,L,groups){
I<-groups;p<-length(I)
SVD<-H.SVD(Y,L)
LambdaI<-matrix(diag(SVD$d)[I,I],p,p)
SVD$u[,I]%*%LambdaI%*%t(SVD$v[,I])
}
```

Note that the trajectory matrix in HMSSA is a block Hankel matrix; thus, we can use the R function `DiagAver()` that we previously provided for univariate SSA by a loop to construct the block Hankel matrix. This function is referred to as `HDiagAver()`:

Program 2.4 Block Horizontal Hankelization R code

```
HDiagAver<-function(X,series.lengths){
M<-length(series.lengths)
L<-nrow(X);k<-series.lengths-L+1
cols<-cumsum(c(0,k))
D<-NULL
for(j in 1:(M)){
D[[j]]<-DiagAver(X[,((cols[j]+1):cols[(j+1)])])}
D
}
```

Applying window length $L = 3$ and group $I = \{1, 2\}$ in Example 2.2, we have:

```
xhat<-H.Group(Y,3,c(1,2))
HDiagAver(xhat,c(6,6))
[[1]]
[1] 1.041995 1.912851 2.940066 3.920088 5.185260
[6] 5.878263

[[2]]
[1] -0.8635822  0.3333589  2.7397766  5.2490031
[5]  6.7763030  2.1151951
```

Applying the above functions: `HMSSA.Hankel()`, `H.SVD()`, `H.Group()` and `HDiagAver()`, it is possible to write a general function to calculate the components of a time series by HMSSA. In this case the function is called HSSA.Rec and is defined as follows:

Program 2.5 HMSSA reconstruction R code

```
HMSSA.Rec<-function(Y,L,groups){
N<-NULL
if(is.list(Y)){
M<-length(Y)
for(i in 1:M)
N[i]<-length(Y[[i]])}
else{
Y<-as.matrix(Y)
```

```
M<-ncol(Y)
N<-rep(nrow(Y),M)}
X<-HMSSA.Hankel(Y,L)
XI<-H.Group(X,L,groups)
Approx<-HDiagAver(XI,N)
if(is.list(Y)){
Resid<-vector("list",M)
for(i in 1:M)
Resid[[i]]<-Y[[i]]-Approx[[i]]}
else{
Approx<-matrix(unlist(Approx),ncol=M)
Resid<-Y-Approx}
list(Approximation=Approx,Residual=Resid)}
```

Similar to the univariate case, we compute both the reconstructed components and residuals by the function HMSSA.Rec(). Let us give an example to see the output with the R function HMSSA.Rec().

Example 2.3: Filtering a real data by HMSSA

The R package Mcomp includes M1 data sets that one can use to see the results of time series analysis. Here, we have extracted the data pertaining to the total number of annual new registrations of motor vehicles for UK, France and Germany from 1969 to 1986. The following R code can be used to filter with HMSSA:

```
install.packages("Mcomp")
require(Mcomp)
FR<-M1$YAI1$x
GR<-M1$YAI12$x
UK<-M1$YAI17$x
motor<-cbind(UK,FR,GR)
```

By calling motor, users can then view the observations on the time series.

```
motor
Time Series:
Start = 1969
End = 1986
Frequency = 1
```

	UK	FR	GR
1969	246.8	67.5	7.3
1970	299.6	62.3	6.3
1971	246.4	112.4	26.8
1972	289.8	175.5	114.5
1973	250.9	238.8	262.7
1974	246.9	317.9	253.0
1975	293.5	378.5	276.8
1976	417.4	378.0	320.1
1977	525.1	443.4	466.3
1978	689.9	541.2	599.2
1979	582.6	608.1	683.5
1980	595.7	639.2	720.5
1981	758.7	697.4	864.0
1982	871.1	664.4	1011.9
1983	1074.4	761.7	1162.4
1984	1008.0	838.0	1308.7
1985	1020.3	1044.9	1419.5
1986	1272.2	1191.0	1472.2

In order to analyse motor by the first three eigentriples of the trajectory matrix and window length $L = 9$, use the following R codes:

```
Approx<-HMSSA.Rec(motor,9,c(1:3))$Approximation
round(Approx,1)
         [,1]    [,2]    [,3]
 [1,]   203.7    59.6   -12.8
 [2,]   223.1    79.2    -2.3
 [3,]   255.0   118.1    47.5
 [4,]   281.9   178.4   130.0
 [5,]   279.2   248.3   207.1
 [6,]   282.5   311.3   252.3
 [7,]   327.1   358.4   293.0
 [8,]   422.1   399.1   360.8
 [9,]   522.6   451.3   462.7
[10,]   590.4   517.7   579.1
[11,]   612.8   590.4   669.5
[12,]   645.5   644.4   747.3
[13,]   747.0   675.9   855.9
[14,]   891.3   702.3  1004.6
```

[15,] 1013.7   768.1 1164.8
[16,] 1052.8   876.6 1300.2
[17,] 1069.4 1021.1 1404.5
[18,] 1174.4 1156.6 1513.3

The above matrix is an approximation of the original trajectory matrix constructed by the total number of annual new registrations of motor vehicles for UK, France and Germany from 1969 to 1986. For example, the first column of the reconstructed matrix, using the first three eigentriples, corresponds to the UK time series. Note also that negative number in the third column indicates that the selection of L and r are not appropriate, and other selections should be considered. Next, we will show how the above matrix can be approximated or filtered using the HMSSA approach.

2.2.2 MSSA: Vertical Form (VMSSA)

In this subsection, we describe the vertical form of MSSA and contrast it with the horizontal form described in Sect. 2.2.1. The time series framework is as for HMSSA, however, here we consider M time series with different series lengths, N_i.

Stage I. Decomposition

Step 1: Embedding. The result of the embedding step is a trajectory matrix $\mathbf{X}^{(i)} = [X_1^{(i)}, \ldots, X_{K_i}^{(i)}] = (x_{mn})_{m,n=1}^{L_i, K_i}$. In VMSSA, the trajectory matrix $\mathbf{X}^{(i)}$ is a Hankel matrix; thus, the embedding procedure for each series separately provides M different $L_i \times K_i$ trajectory matrices $\mathbf{X}^{(i)}$ ($i = 1, \ldots, M$). To form a new block Hankel matrix in a vertical form, it is necessary to have $K_1 = \ldots, K_M = K$. Accordingly, this version of MSSA enables various window lengths L_i and different series lengths N_i; however, K_i is the same for all series. Denote by $\mathbf{X}^{(i)}$ the trajectory matrix associated with time series i with $i = 1, \ldots, M$, the result of this step is a vertically stacked block Hankel trajectory matrix as follows:

$$\mathbf{X}_V = \begin{bmatrix} \mathbf{X}^{(1)} \\ \vdots \\ \mathbf{X}^{(M)} \end{bmatrix},$$

where $\mathbf{X}_V$ indicates that the output of the first step is a block Hankel trajectory matrix in a *vertical* form.

Step 2: SVD. In this step, the SVD of $\mathbf{X}_V$ is performed. Denote by $\lambda_{V_1}, \ldots, \lambda_{V_{Lsum}}$ the eigenvalues of $\mathbf{X}_V \mathbf{X}_V^T$, arranged in decreasing order $(\lambda_{V_1} \geq \cdots \lambda_{V_{Lsum}} \geq 0)$ and by $U_{V_1}, \ldots, U_{V_{Lsum}}$, the corresponding eigenvectors, and $L_{sum} = \sum_{i=1}^{M} L_i$. Note that the structure of the matrix $\mathbf{X}_V \mathbf{X}_V^T$ is as follows:

$$\mathbf{X}_V \mathbf{X}_V^T = \begin{bmatrix} \mathbf{X}^{(1)} \mathbf{X}^{(1)T} & \mathbf{X}^{(1)} \mathbf{X}^{(2)T} & \cdots & \mathbf{X}^{(1)} \mathbf{X}^{(M)T} \\ \mathbf{X}^{(2)} \mathbf{X}^{(1)T} & \mathbf{X}^{(2)} \mathbf{X}^{(2)T} & \cdots & \mathbf{X}^{(2)} \mathbf{X}^{(M)T} \\ \vdots & \vdots & \ddots & \vdots \\ \mathbf{X}^{(M)} \mathbf{X}^{(1)T} & \mathbf{X}^{(M)} \mathbf{X}^{(2)T} & \cdots & \mathbf{X}^{(M)} \mathbf{X}^{(M)T} \end{bmatrix}. \quad (2.5)$$

The structure of the matrix $\mathbf{X}_V \mathbf{X}_V^T$ is similar to the variance-covariance matrix in classical multivariate statistical analysis literature. The matrix $\mathbf{X}^{(i)} \mathbf{X}^{(i)T}$, which is used in univariate SSA, for the series $Y_{N_i}^{(i)}$, appears along the diagonal, and the products of two Hankel matrices $\mathbf{X}^{(i)} \mathbf{X}^{(j)T}$ ($i \neq j$), which are related to the series $Y_{N_i}^{(i)}$ and $Y_{N_j}^{(j)}$, appear in the off-diagonal.

The SVD of $\mathbf{X}_V$ can be written as $\mathbf{X}_V = \mathbf{X}_{V_1} + \cdots + \mathbf{X}_{V_{Lsum}}$, where

$$\mathbf{X}_{V_i} = \sqrt{\lambda_i} U_{V_i} V_{V_i}^T, \quad (2.6)$$

$$V_{V_i} = \mathbf{X}_V^T U_{V_i} / \sqrt{\lambda_{V_i}} \ (\mathbf{X}_{V_i} = 0 \text{ if } \lambda_{V_i} = 0). \quad (2.7)$$

Stage II. Reconstruction

Step 1: Grouping. This step corresponds to splitting the matrices $\mathbf{X}_{V_1}, \ldots, \mathbf{X}_{V_{Lsum}}$ into several disjoint groups and summing the matrices within each group. The split of the set of indices $\{1, \ldots, L_{sum}\}$ into disjoint subsets $I_1, \ldots, I_m$ corresponds to the representation $\mathbf{X}_V = \mathbf{X}_{I_1} + \cdots + \mathbf{X}_{I_m}$. As in SSA and HMSSA, the procedure of choosing the sets $I_1, \ldots, I_m$ is called grouping. For a given group I, the contribution of the component $\mathbf{X}_{V_I}$ is measured by the share of the corresponding eigenvalues: $\sum_{i \in I} \lambda_{V_i} / \sum_{i=1}^{d_V} \lambda_{V_i}$, where d_V is the rank of $\mathbf{X}_V$ and $I \subset \{1, \ldots, L_{sum}\}$. As a simple case where there are only signal and noise components, two groups of indices are used, $I_1 = \{1, \ldots, r\}$ and $I_2 = \{r+1, \ldots, L_{sum}\}$ the group $I = I_1$ is associated with the signal component and the group I_2 with noise.

Step 2: Diagonal averaging or Hankelization. The purpose of diagonal averaging is to transform the reconstructed matrix $\widehat{\mathbf{X}}_{V_i}$, corresponding to each series $Y_{N_i}^{(i)}$, to the form of a Hankel matrix, which can be subsequently converted to a time series. Let $\widetilde{\mathbf{X}}^{(i)}$ be the approximation of $\mathbf{X}^{(i)}$ obtained from the diagonal averaging step. If $\widetilde{x}_{mn}^{(i)}$ stands for an element of a matrix $\widetilde{\mathbf{X}}^{(i)}$, then the jth term of the reconstructed series $\widetilde{Y}_{N_i}^{(i)} = (\widetilde{y}_1^{(i)}, \ldots, \widetilde{y}_j^{(i)}, \ldots, \widetilde{y}_{N_i}^{(i)})$ is obtained by arithmetically averaging $\widetilde{x}_{mn}^{(i)}$ over all (m, n), such that $m + n - 1 = j$.

R codes for the VMSSA approach are provided below, similar to HMSSA, where there are four functions VMSSA.Hankel(), V.SVD(), V.Group() and VDiagAver() in order to carry out the four steps in VMSSA filtering algorithm

Program 2.6 VMSSA Trajectory matrix construction R code

```
VMSSA.Hankel<-function(Y,k){
X<-NULL
if(is.list(Y)){
M<-length(Y)
for(i in 1:M){
Li<-length(Y[[i]])-k+1
Xi<-outer((1:Li),(1:k),function(x,y) Y[[i]][(x+y-1)])
X<-rbind(X,Xi)}
}
else{
Y<-as.matrix(Y)
M<-ncol(Y)
for(i in 1:M){
Li<-length(Y[,i])-k+1
Xi<-outer((1:Li),(1:k),function(x,y) Y[(x+y-1),i])
X<-rbind(X,Xi)}
}
X
}
```

Program 2.7 VMSSA SVD R code

```
V.SVD<-function(Y,k){
X<-VMSSA.Hankel(Y,k)
svd(X)
}
```

Program 2.8 VMSSA grouping R code

```
V.Group<-function(Y,k,groups){
I<-groups;p<-length(I)
SVD<-V.SVD(Y,k)
LambdaI<-matrix(diag(SVD$d)[I,I],p,p)
SVD$u[,I]%*%LambdaI%*%t(SVD$v[,I])
}
```

Program 2.9 VMSSA hankelization R code

```
VDiagAver<-function(X,series.lengths){
M<-length(series.lengths)
k<-ncol(X);L<-series.lengths-k+1
rows=cumsum(c(0,L))
D<-NULL
for(j in 1:(M)){
D[[j]]<-DiagAver(X[((rows[j]+1):rows[(j+1)]),])}
D
}
```

Applying the above functions, we can write a new R function, VMSSA.Rec, to obtain the reconstructed components and residuals.

Program 2.10 VMSSA reconstruction R code

```
VMSSA.Rec<-function(Y,k,groups){
N<-NULL
if(is.list(Y)){
M<-length(Y)
for(i in 1:M)
N[i]<-length(Y[[i]])}
else{
Y<-as.matrix(Y)
M<-ncol(Y)
N<-rep(nrow(Y),M)}
X<-VMSSA.Hankel(Y,k)
XI<-V.Group(Y,k,groups)
Approx<-VDiagAver(XI,N)
if(is.list(Y)){
Resid<-vector("list",M)
for(i in 1:M)
```

```
Resid[[i]]<-Y[[i]]-Approx[[i]]}
else{
Approx<-matrix(unlist(Approx),ncol=M)
Resid<-Y-Approx}
list(Approximation=Approx,Residual=Resid)}
```

The results of using this function with Example 2.3 are as follows, where we have again considered the first three eigentriples and $L_i = L = 9$, which corresponds to $k = 10$.

```
Approx<-VMSSA.Rec(motor,10,c(1:3))$Approximation
round(Approx,1)
         [,1]    [,2]    [,3]
 [1,]   170.1    60.6   -16.9
 [2,]   194.8    88.9     2.6
 [3,]   234.3   131.4    56.7
 [4,]   273.2   185.1   133.2
 [5,]   287.3   247.5   200.5
 [6,]   294.4   306.4   244.3
 [7,]   338.2   354.3   288.6
 [8,]   427.3   397.0   363.1
 [9,]   523.5   448.1   469.9
[10,]   587.6   517.5   578.6
[11,]   608.4   591.3   670.4
[12,]   642.2   648.8   746.9
[13,]   744.0   682.1   852.2
[14,]   889.8   707.5   999.4
[15,]  1014.0   767.4  1164.2
[16,]  1055.4   869.6  1304.2
[17,]  1073.1  1012.4  1410.9
[18,]  1177.3  1155.2  1513.8
```

Similar to VMSSA case, the above matrix is an approximation of the original trajectory matrix constructed by the total number of annual new registrations of motor vehicles for UK, France and Germany using HMSSA approach. Similarly, for instance, the first column of the reconstructed matrix, using the first three eigentriples, corresponds to the UK time series. A simple comparison between two filtered/approximated matrices obtained by VMSSA and HMSSA indicates the difference between two filtered series of all countries used in this example. Note that a negative number in third column shows that proper eigentriples should be considered. This issue will be discussed in the next subsection.

2.3 Choosing Parameters in MSSA

The criteria for parameter selection in MSSA are yet to be theoretically determined as in its univariate counterpart, SSA. In this section, we discuss the problem further and in doing so we mainly follow Hassani and Mahmoudvand (2013). The situation is more complicated with MSSA compared to univariate SSA, in which similarity and orthogonality among series play an important role for selecting L and r. Moreover, the multivariate case deals with a block trajectory Hankel matrix with special features rather than one Hankel matrix. This makes the problem even more complex. The first step of the SSA algorithm provides a Hankel trajectory matrix, which plays an important role in the SSA technique as the other steps depend on its structure and eigenvalues extracted obtained from this matrix. In MSSA this matrix can be formed differently, which will change the forecasting results significantly in some situations. Furthermore, depending on the structure of this matrix, different forecasting algorithms can be formed. This matrix also depends on the window length L. Those interested in a detailed study of the characteristics of the trajectory matrix with respect to different values of the L are referred to Hassani and Mahmoudvand (2013).

Note also that the second SSA choice, which is the number of eigenvalues r, is very important for the reconstruction stage. The aim in this selection is to provide the noise-free series, which is selected for forecasting. The selection of r is difficult for the multivariate case, as each eigenvalue contains information from all the series considered in the multivariate analysis. Similar to other related multivariate analysis, it is also important to know how different time series change in relation to each other, or how they are associated. To evaluate this, the trajectory matrix, which is a block Hankel matrix, needs to be considered. For the SSA, orthogonality among the reconstructed series that are obtained from the extracted eigenvalues from the trajectory matrix, is foremost; therefore, the number of common/-matched components among time series needs to be considered. In Hassani and Mahmoudvand (2013), the authors evaluate the effect of matched components among series from a theoretical aspect and also through comprehensive simulation studies, and show similarity and dissimilarity among series affect the forecasting performance. We define some criteria, as with univariate SSA, to guide the selection of parameters for multivariate SSA.

2.3.1 Window Length(s)

First, recall that the criteria considered in Chapter 1 suggested that the optimal value of the window length parameter, at least for the reconstruction stage, is close to the Median$\{1, \ldots, N\}$, although this bound may not provide the optimal forecasting result. Below, we will see that the quantiles of the number set $\{1, \ldots, N\}$ can be good choices for window length in MSSA.

Rank of the Trajectory Matrix

Considering the trajectory matrix $\mathbf{X}$ of dimension $LM \times K$ for VMSSA, or $K \times ML$ for HMSSA (assuming all series have identical series length and we have chosen similar L), Theorem B.2 shows that the maximum rank of the trajectory matrix, which corresponds to the maximum number of extracted components of the series, is attained when:

$$L = \begin{cases} \left\lceil \frac{1}{M+1} \right\rceil (N+1) & \text{VMSSA;} \\ \left\lceil \frac{M}{M+1} \right\rceil (N+1) & \text{HMSSA,} \end{cases} \tag{2.8}$$

where the notation $\lceil u \rceil$ denotes the closest integer number to u, for all u.

The Lag-Covariance Matrices

We consider the behaviour of the matrix $\mathbf{XX}^T/K$, which can be viewed as the lag-covariance matrix. Note that K indicates the number of lagged vectors as in classical multivariate analysis. Denote by $T_\mathbf{X}^{L,N}$ the trace of the matrix $\mathbf{XX}^T$, $tr(\mathbf{XX}^T)$, Theorem B.5 shows that $T_\mathbf{X}^{L,N}/K$ is an increasing function of L and therefore the window length which is given by Eq. (2.8) can be considered as a suitable value for reconstruction of the series.

Recall again that SSA is a technique that converts a time series into a set of interpretable components. This transformation is defined in such a way that the first component has the largest possible variance (i.e. it accounts for as much of the variability in the series as possible), and each succeeding component in turn has the highest variance possible under the constraint that it is well separated from the preceding components. Therefore, the window length provided by Eq. (2.8) produces the highest total variability.

2.3.2 Grouping Parameter, r

We consider the issues of orthogonality and similarity (with respect to some criteria) between two matrices $\mathbf{X}^{(i)}$ and $\mathbf{X}^{(j)}$ ($i \neq j$) which plays an important role in MSSA. Consider two trajectory Hankel matrices $\mathbf{X}^{(i)}$ and $\mathbf{X}^{(j)}$ ($i \neq j$) corresponding to the series $Y_N^{(i)}$ and $Y_N^{(j)}$, respectively. Then, the matrices $\mathbf{X}^{(i)}$ and $\mathbf{X}^{(j)}$ ($i \neq j$) are orthogonal if

$$\mathbf{X}^{(i)}\mathbf{X}^{(j)^T} = 0, \quad \mathbf{X}^{(i)^T}\mathbf{X}^{(j)} = 0. \tag{2.9}$$

In this case, the structure of matrix $\mathbf{X}_V \mathbf{X}_V^T$, the vertical form, see Eq. (2.5), is simplified as follows:

$$\mathbf{X}_V \mathbf{X}_V^T = \begin{bmatrix} \mathbf{X}^{(1)}\mathbf{X}^{(1)T} & 0 & \cdots & 0 \\ 0 & \mathbf{X}^{(2)}\mathbf{X}^{(2)T} & \cdots & 0 \\ \vdots & \vdots & \ddots & \vdots \\ 0 & 0 & \cdots & \mathbf{X}^{(M)}\mathbf{X}^{(M)T} \end{bmatrix}. \tag{2.10}$$

In this case, we are dealing with a diagonal matrix and there is no cross product between trajectory matrices $\mathbf{X}^{(i)}$ and $\mathbf{X}^{(j)}$ in the off-diagonal. Therefore, the total variation of $\mathbf{X}_V$ is distributed among matrices $\{\mathbf{X}^{(1)}, \ldots, \mathbf{X}^{(M)}\}$. Furthermore, we have

$$r = r_1 + r_2 + \cdots + r_M, \tag{2.11}$$

where $r_i = \text{rank}\,\mathbf{X}^{(i)}$ ($i = 1, \ldots, M$) is the trajectory dimension in univariate SSA for each series separately, and $r = \text{rank}\,\mathbf{X}_V$ is the trajectory dimension in multivariate SSA for all series. It should be mentioned that r_i is called the number of singular values, used in the reconstruction stage, in SSA terminology. In general, we have

$$r \leq r_1 + r_2 + \cdots + r_M. \tag{2.12}$$

The equality holds only in the orthogonality situation, see Eq. (2.9). Note also that in the case of orthogonality, the structure of the matrix $\mathbf{X}_H \mathbf{X}_H^T$ for the HMSSA version is as follows:

$$\mathbf{X}_H \mathbf{X}_H^T = \mathbf{X}^{(1)}\mathbf{X}^{(1)^T} + \cdots + \mathbf{X}^{(M)}\mathbf{X}^{(M)^T}, \tag{2.13}$$

which is the same as for the general case in Eq. (2.4). Thus, orthogonality does not change the structure of $\mathbf{X}_H\mathbf{X}_H^T$ and the equality $r = r_1 + r_2 + \cdots + r_M$ holds for HMSSA.

Obviously, we do not expect a feedback system if the series are orthogonal. In a feedback system, the multivariate forecasting approach improves the forecasting results of all M series. For example, in a feedback system with two series, $M = 2$, two series are mutually supportive. If the M series are similar according to some criteria, then $r = r_1 = \cdots = r_M$. However, there are several cases where one series is more supportive than the other one. In this case, MSSA provides better results, at least for one of the series, as there are some matched components between series. If all components are matched, then we have similar series, and if there is not any matched components, we then have the issue of orthogonality.

We can suggest the following bound for the number of singular values r for the VMSSA case:

$$1 \leq R_m \leq r \leq R_m + R_{um} \leq ML, \qquad (2.14)$$

where R_m and R_{um} denote the number of matched (at least one) and unmatched components in the whole multivariate system, respectively. Note that for HMSSA, the upper bound is L. The lower bound in Eq. (2.14) is suitable if the aim is forecasting the series that contains the minimum component, using information of the other series, in the whole system. If the aim is forecasting the series with maximum components, then the suitable lower bound is $\text{Max}\{r_i\}_{i=1}^M$. In both cases, the lower bounds occur when all series are exactly similar. Furthermore, r takes the upper bound for the case of full orthogonality.

The implication of this discussion is that the ceiling to the multivariate trajectory dimension is simply that obtained when there are, in a sense, no common or matched components among the M series; the presence of matching components reduces the dimension of the multivariate system and indicates that the system is interrelated, which should improve the forecast. The selection of $r < R_m$ leads to a loss of precision as parts of the signals in all series will be lost. On the other hand, if $r > R_m + R_{um}$, then noise is included in the reconstructed series. The selection of $r \cong R_m$ (keeping $r > R_m$) is a good choice for highly interrelated series sharing several common components. The selection of $r \cong R_m + R_{um}$ is necessary when the series analysed have very little relation to each other.

2.4 Forecasting by MSSA

The univariate SSA forecasting algorithm is based on the two main forecasting approaches referred to in Sects. 1.4.1 and 1.4.2 as recurrent and vector forecasting, that is RSSA and VSSA, respectively. Considering the recurrent and vector approaches, there are four different forecasting algorithms for MSSA as follows:

$$\text{MSSA forecasting approach} = \begin{cases} \text{HMSSA} \begin{cases} \text{Recurrent approach} \\ \text{Vector approach} \end{cases} \\ \text{VMSSA} \begin{cases} \text{Recurrent approach} \\ \text{Vector approach} \end{cases} \end{cases}$$

2.4.1 HMSSA Recurrent Forecasting Algorithm (HMSSA-R)

1. For a fixed value of L, construct the trajectory matrix $\mathbf{X}^{(i)} = [X_1^{(i)}, \ldots, X_K^{(i)}] = (x_{mn})_{m,n=1}^{L,K_i}$ for each single series $Y_{N_i}^{(i)}$ ($i = 1, \ldots, M$) separately.
2. Construct the block trajectory matrix $\mathbf{X}_H$ as follows:

$$\mathbf{X}_H = [\mathbf{X}^{(1)} : \mathbf{X}^{(2)} : \ldots : \mathbf{X}^{(M)}].$$

3. Define $U_{H_j} = (u_{1j}, \ldots, u_{Lj})^T$, with length L, as the jth eigenvector of $\mathbf{X}_H \mathbf{X}_H^T$.
4. Where $\widehat{\mathbf{X}}$ is a Hankel operator, consider $\widehat{\mathbf{X}}_H = \sum_{i=1}^{r} U_{H_i} U_{H_i}^T \mathbf{X}_H$ which is the reconstructed matrix obtained using r eigentriples:

$$\mathbf{X}_H = \left[\widehat{\mathbf{X}}^{(1)} : \widehat{\mathbf{X}}^{(2)} : \ldots : \widehat{\mathbf{X}}^{(M)}\right].$$

5. Consider matrix $\widetilde{\mathbf{X}}^{(i)} = \mathscr{H}\widehat{\mathbf{X}}^{(i)}$ ($i = 1, \ldots, M$) as the result of the Hankelization procedure of the matrix $\widehat{\mathbf{X}}^{(i)}$ obtained from the previous step.
6. Let $U_{H_j}^{\triangledown}$ denote the vector of the first $L-1$ coordinates of the eigenvectors U_{H_j}, and π_{H_j} indicate the last coordinate of the eigenvectors U_{H_j} ($j = 1, \ldots, r$).

7. Define $v^2 = \sum_{j=1}^{r} \pi_{H_j}^2$.
8. Denote the linear coefficients vector $\mathscr{R}$ as follows:

$$\mathscr{R} = \frac{1}{1-v^2} \sum_{j=1}^{r} \pi_{H_j} U_{H_j}^{\nabla}. \qquad (2.15)$$

9. If $v^2 < 1$, then the h-step ahead HMSSA forecasts exist and are calculated by the following formula:

$$\left[\hat{y}_{j_i}^{(1)}, \ldots, \hat{y}_{j_M}^{(M)}\right] = \begin{cases} \left[\tilde{y}_{j_1}^{(1)}, \ldots, \tilde{y}_{j_M}^{(M)}\right], & j_i = 1, \ldots, N_i, \\ \mathscr{R}^T \mathbf{Z}_h, & j_i = N_i + 1, \ldots, N_i + h, \end{cases} \qquad (2.16)$$

where $\mathbf{Z}_h = \left[Z_h^{(1)}, \ldots, Z_h^{(M)}\right]$ and $Z_h^{(i)} = \left[\hat{y}_{N_i-L+h+1}^{(i)}, \ldots, \hat{y}_{N_i+h-1}^{(i)}\right]^T$ ($i = 1, \ldots, M$).

Recall that $\tilde{y}_{j_i}^{(m)}$ is the j_ith observation of the reconstructed series m, where $m = 1, \ldots, M$.

Note that formula (2.16) indicates that the h-step ahead forecasts of each series are achieved by the same LRF generated considering all series in a multivariate system. To compare the approaches of the VMSSA and HMSSA forecasting algorithms, we consider, similarities and dissimilarities of these approaches from different perspectives. The series length, the value of window length (L_i), the number of nonzero singular values obtained from the block trajectory matrix and the LRF. Table 2.1 provides a summary of the comparison.

Table 2.1 Similarities and dissimilarities between the VMSSA and HMSSA recurrent forecasting algorithms

Method	Series length	L_i	K_i	Number of the eigenvalues	LRF
VMSSA	Different	Different	Equal	$\sum L_i$	Different
HMSSA	Different	Equal	Different	L	Equal

R Code for HMSSA-R

Program 2.11 provides the HMSSA-R algorithm. First, recall that in order to run this function, you must first define functions `HMSSA.Hankel()`, `H.SVD()`, `H.Group()`, `HDiagAver()` and `HMSSA.Rec()`, in R.

Program 2.11 Recurrent HMSSA forecasting R code

```
HMSSA.R<-function(L,Group,h,Y){
r<-length(Group)
Dec<-H.SVD(Y,L)
sigma<-Dec$d
d<-length(sigma[sigma>0])
U<-matrix(Dec$u,L,d)
V<-matrix(Dec$v,ncol=d)
hx<-HMSSA.Rec(Y,L,Group)$Approximation
v<-array(U[L,Group],dim=c(1,r))
Ud<-array(U[-L,Group],dim=c((L-1),r))
R<-1/(1-sum(v^2))*v%*%t(Ud)
if(is.list(Y)){
M<-length(Y)
forecasts<-vector("list",M)
for(i in 1:M){
N<-length(Y[[i]])
for(j in 1:h){
Z<-array(hx[[i]][(N-L+1+j):(N+j-1)],dim=c(L-1))
forecasts[[i]]<-c(forecasts[[i]],R%*%Z)
hx[[i]]<-c(hx[[i]],R%*%Z)}}
matrix(unlist(forecasts),ncol=M)}
else{
N<-nrow(Y);M<-ncol(Y)
for(j in 1:h){
Z<-array(hx[(N-L+1+j):(N+j-1),],dim=c((L-1),M))
hx<-rbind(hx,R%*%Z)}
hx[(N+1):(N+h),]}}
```

> **Example 2.4: Recurrent HMSSA forecasting for the motor data set**

Assume we would like to obtain forecasts for 1987–1992 in Example 2.3, that is annual new registration cars in the UK, France and Germany. Applying the window length $L = 9$ and first three eigentriples for reconstruction, we find that:

```
round(HMSSA.R(9,c(1:3),6,motor),1)
     [,1]    [,2]    [,3]
[1,] 1313.2  1234.4  1694.0
[2,] 1503.2  1322.8  1908.5
[3,] 1673.5  1440.7  2145.8
[4,] 1796.0  1610.0  2371.3
[5,] 1908.5  1813.3  2582.7
[6,] 2072.8  2014.7  2814.1
```

You can also see the real observations by:

```
FR.new<-M1$YAI1$xx
GR.new<-M1$YAI12$xx
UK.new<-M1$YAI17$xx
motor.new<-cbind(UK.new,FR.new,GR.new)
motor.new
Time Series:
Start = 1987
End = 1992
Frequency = 1
     UK.new  FR.new  GR.new
1987 1481.4  1211.6  1554.10
1988 1424.3  1222.8  1729.60
1989 1357.1  1367.8  1709.10
1990 1405.8  1382.7  1535.30
1991 1389.1  1406.7  1607.30
1992 1265.6  1589.0  2058.19
```

It is worth mentioning that the above forecasting results shown here are for illustration purpose only. One should chose proper number of eigenvalues to obtain accurate forecast. The selection of the eigentriples in MSSA will be discussed later.

2.4.2 VMSSA Recurrent Forecasting Algorithm (VMSSA-R)

Consider M series $Y_{N_i}^{(i)} = (y_1^{(i)}, \ldots, y_{N_i}^{(i)})$ and the corresponding window lengths L_i, $2 \leq L_i \leq N_i - 1$, $i = 1\ldots, M$. The VMSSA forecasting algorithm for the h-step ahead forecast is as follows.

1. For a fixed value of K, construct the trajectory matrix $\mathbf{X}^{(i)} = [X_1^{(i)}, \ldots, X_K^{(i)}] = (x_{mn})_{m,n=1}^{L_i,K}$ for each single series $Y_{N_i}^{(i)}$ ($i = 1, \ldots, M$) separately.
2. Construct the block trajectory matrix $\mathbf{X}_V$ as follows:

$$\mathbf{X}_V = \begin{bmatrix} \mathbf{X}^{(1)} \\ \vdots \\ \mathbf{X}^{(M)} \end{bmatrix}.$$

3. Let $\mathbf{U}_{V_j} = (U_j^{(1)}, \ldots, U_j^{(M)})^T$ be the jth eigenvector of the $\mathbf{X}_V \mathbf{X}_V^T$, where $U_j^{(i)}$, with length L_i, corresponds to the series $Y_{N_i}^{(i)}$ ($i = 1, \ldots, M$).
4. Let $\widehat{\mathbf{X}}_V = [\widehat{X}_1 : \ldots : \widehat{X}_K] = \sum_{i=1}^r U_{V_i} U_{V_i}^T \mathbf{X}_V$ denote the reconstructed matrix achieved from r eigentriples, that is:

$$\widehat{\mathbf{X}}_V = \begin{bmatrix} \widehat{\mathbf{X}}^{(1)} \\ \vdots \\ \widehat{\mathbf{X}}^{(M)} \end{bmatrix}.$$

5. Consider matrix $\widetilde{\mathbf{X}}^{(i)} = \mathscr{H} \widehat{\mathbf{X}}^{(i)}$ ($i = 1, \ldots, M$) as the result of the Hankelization procedure of the matrix $\widehat{\mathbf{X}}^{(i)}$ obtained from the previous step.
6. Let $U_j^{(i)\nabla}$ denote the vector of the first $L_i - 1$ components of the vector $U_j^{(i)}$ and $\pi_j^{(i)}$ the last component of the vector $U_j^{(i)}$ ($i = 1, \ldots, M$).
7. Select the number of r eigentriples for the reconstruction stage that will also be used for forecasting.
8. Define the matrix $\mathbf{U}^{\nabla M} = (U_1^{\nabla M}, \ldots, U_r^{\nabla M})$, where $U_j^{\nabla M}$ is as follows:

$$U_j^{\nabla M} = \begin{bmatrix} U_j^{(1)\nabla} \\ \vdots \\ U_j^{(M)\nabla} \end{bmatrix}.$$

9. Define the matrix $\mathbf{W}$ as follows:

$$\mathbf{W} = \begin{bmatrix} \pi_1^{(1)} & \pi_2^{(1)} & \cdots & \pi_r^{(1)} \\ \pi_1^{(2)} & \pi_2^{(2)} & \cdots & \pi_r^{(2)} \\ \vdots & \vdots & \cdots & \vdots \\ \pi_1^{(M)} & \pi_2^{(M)} & \cdots & \pi_r^{(M)} \end{bmatrix}.$$

10. If the matrix $(\mathbf{I}_{M \times M} - \mathbf{W}\mathbf{W}^T)^{-1}$ exists and $r \leq L_{sum} - M$, then the h-step ahead VMSSA forecasts exist and are obtained by the following formula:

$$\begin{bmatrix} \hat{y}_{j_1}^{(1)} \\ \vdots \\ \hat{y}_{j_M}^{(M)} \end{bmatrix} = \begin{cases} \left[\tilde{y}_{j_1}^{(1)}, \ldots, \tilde{y}_{j_M}^{(M)}\right]^T, & j_i = 1, \ldots, N_i \\ \left(\mathbf{I}_{M \times M} - \mathbf{W}\mathbf{W}^T\right)^{-1} \mathbf{W}\mathbf{U}^{\triangledown M^T} \mathbf{Z}_h, & j_i = N_i + 1, \ldots, N_i + h, \end{cases}$$
(2.17)

where $\mathbf{Z}_h = \left[Z_h^{(1)}, \ldots, Z_h^{(M)}\right]^T$ and $Z_h^{(i)} = \left[\hat{y}_{N_i - L_i + h + 1}^{(i)}, \ldots, \hat{y}_{N_i + h - 1}^{(i)}\right]$ ($i = 1, \ldots, M$). It should be noted that formula (2.17) indicates that the h-step ahead forecasts of the less noisy series $\hat{Y}_{N_i}^{(i)}$ are obtained by a multidimensional linear recurrent formula (LRF) where, as in the univariate case, there is only a one-dimensional LRF.

R Codes for VMSSA-R

Program 2.12 produces VMSSA-R forecasts. In order to run this function, you have to call functions VMSSA.Hankel(), V.SVD(), V.Group(), VDiagAver() and VMSSA.Rec() in R.

Program 2.12 Recurrent VMSSA forecasting R code

```
VMSSA.R<-function(k,Group,h,Y){
r<-length(Group)
Dec<-V.SVD(Y,k)
sigma<-Dec$d
d<-length(sigma[sigma>0])
U<-matrix(Dec$u,ncol=d)
V<-matrix(Dec$v,d,k)
hx<-VMSSA.Rec(Y,k,Group)$Approximation
if(is.list(Y)){
M<-length(Y);L<-NULL
forecasts<-array(0,dim=c(h,M))
for(i in 1:M) L[i]<-length(Y[[i]])-k+1
```

```
Ld<-c(0,cumsum(L-1))
W<-array(U[cumsum(L),Group],dim=c(M,r))
Ud<-array(U[-cumsum(L),Group],dim=c((sum(L)-M),r))
R<-solve(diag(M)-W%*%t(W))%*%W%*%t(Ud)
for(i in 1:h){
for(j in 1:M){
for(m in 1:M){
Zh<-hx[[m]][(k+i):(k+L[m]+i-2)]
jm<-(Ld[m]+1):Ld[(m+1)]
forecasts[i,j]<-forecasts[i,j]+R[j,jm]%*%Zh}
hx[[j]]<-c(hx[[j]],forecasts[i,j])}}
forecasts}
else{
M<-ncol(Y);L<-rep((nrow(Y)-k+1),M)
forecasts<-array(0,dim=c(h,M))
W<-array(U[cumsum(L),Group],dim=c(M,r))
Ud<-array(U[-cumsum(L),Group],dim=c((sum(L)-M),r))
R<-solve(diag(M)-W%*%t(W))%*%W%*%t(Ud)
for(i in 1:h){
Zh<-as.vector(hx[(k+i):(k+L[1]+i-2),])
forecasts[i,]<-R%*%Zh
hx<-rbind(hx,forecasts[i,])}
forecasts}}
```

Example 2.5: Recurrent VMSSA forecast for the motor data set

Using $K = 10$ (corresponding to $L = 9$) and the first three eigentriples in Example 2.3 provides the following forecasts:

```
round(VMSSA.R(10,c(1:3),6,motor),1)
       [,1]    [,2]    [,3]
[1,] 1326.7  1197.3  1701.1
[2,] 1499.8  1258.7  1913.4
[3,] 1656.4  1360.0  2122.9
[4,] 1774.9  1514.4  2311.0
[5,] 1883.0  1686.9  2496.5
[6,] 2027.4  1833.1  2714.9
```

2.4.3 HMSSA Vector Forecasting Algorithm (HMSSA-V)

The procedure for HMSSA-V is very similar to its univariate version, VSSA and to HMSSA-R.

HMSSA-V forecasting algorithm

The first parts of this algorithm are as for HMSSA-R. To continue consider the following matrix

$$\Pi = \mathbf{U}^\nabla \mathbf{U}^{\nabla T} + (1 - v^2)\mathscr{R}\mathscr{R}^T,$$

where $\mathbf{U}^\nabla = [U_1^\nabla, \ldots, U_r^\nabla]$. Now consider the linear operator

$$\mathscr{P}^{(v)} : \mathfrak{L}_r \mapsto \mathbb{R}^L,$$

where

$$\mathscr{P}^{(v)} Y = \begin{pmatrix} \Pi Y_\triangle \\ \mathscr{R}^T Y_\triangle \end{pmatrix}, \quad Y \in \mathfrak{L}_r.$$

1. Define the vector $Z_j^{(i)}$ $(i = 1, \ldots, M)$ as follows:

$$Z_j^{(i)} = \begin{cases} \widetilde{X}_j^{(i)} & \text{for } j = 1, \ldots, k_i \\ \mathscr{P}^{(v)} Z_{j-1}^{(i)} & \text{for } j = k_i + 1, \ldots, k_i + h + L - 1 \end{cases} \quad (2.18)$$

where the $\widetilde{X}_j^{(i)}$'s are the reconstructed columns of the trajectory matrix of the ith series after grouping and leaving noise components.

2. By constructing the matrix $\mathbf{Z}^{(i)} = [Z_1^{(i)}, \ldots, Z_{k_i+h+L-1}^{(i)}]$ and performing diagonal averaging we obtain a new series $\hat{y}_1^{(i)}, \ldots, \hat{y}_{N_i+h+L-1}^{(i)}$, where $\hat{y}_{N_i+1}^{(i)}, \ldots, \hat{y}_{N_i+h}^{(i)}$ provides the h-step ahead of HMSSA-V forecast.

R Codes for HMSSA-V

Applying previous R functions, the following R function produces HMSSA-V forecasts:

Program 2.13 Vector HMSSA forecasting R code

```
HMSSA.V<-function(L,Group,h,Y){
r<-length(Group)
```

```
Dec<-H.SVD(Y,L)
sigma<-Dec$d
d<-length(sigma[sigma>0])
U<-matrix(Dec$u,L,d)
V<-matrix(Dec$v,ncol=d)
hx<-H.Group(Y,L,Group)
v<-array(U[L,Group],dim=c(1,r))
Ud<-array(U[-L,Group],dim=c((L-1),r))
R<-1/(1-sum(v^2))*v%*%t(Ud)
pai<-Ud%*%t(Ud)+(1-sum(v^2))*t(R)%*%R
Pv<-rbind(pai,R)
if(is.list(Y)){
M<-length(Y);N<-0
for(i in 1:M) N[i]<-length(Y[[i]])
k<-N-L+1;ks<-c(0,cumsum(k))
forecasts<-array(dim=c(h,M))
for(j in 1:M){
G<-hx[,(ks[j]+1):(ks[(j+1)])]
for(l in (k[j]+1):(N[j]+h))
G<-cbind(G,(Pv%*%G[2:L,(l-1)]))
forecasts[,j]<-DiagAver(G)[(N[j]+1):(N[j]+h)]
}
forecasts
}
else{
M<-ncol(Y);N<-nrow(Y)
k<-N-L+1
forecasts<-array(dim=c(h,M))
for(j in 1:M){
G<-hx[,((j-1)*k+1):(j*k)]
for(l in (k+1):(N+h))
G<-cbind(G,(Pv%*%G[2:L,(l-1)]))
forecasts[,j]<-DiagAver(G)[(N+1):(N+h)]
}
forecasts
}
}
```

Example 2.6: Vector HMSSA forecasting for the motor data set

Using the function HMSSA.V() with $L = 9$ and $I = \{1, 2, 3\}$ for the data in Example 2.3, we obtain the following results:

```
round(HMSSA.V(9,c(1:3),6,motor),1)
       [,1]    [,2]    [,3]
[1,] 1310.9  1203.7  1693.4
[2,] 1431.3  1311.3  1872.2
[3,] 1561.7  1428.1  2065.5
[4,] 1703.2  1554.9  2274.7
[5,] 1856.7  1692.5  2501.2
[6,] 2023.2  1841.9  2746.4
```

2.4.4 VMSSA Vector Forecasting Algorithm (VMSSA-V)

The first 10 parts of this algorithm are as for VMSSA-R, see Sect. 2.4.2. To continue consider the matrix:

$$\boldsymbol{\Pi} = \mathbf{U}^\nabla \mathbf{U}^{\nabla T} + \mathcal{R}\left(\mathbf{I}_{M \times M} - \mathbf{W}\mathbf{W}^T\right)\mathcal{R}^T, \quad (2.19)$$

where $\mathcal{R} = \mathbf{U}^\nabla \mathbf{W}^T \left(\mathbf{I}_{M \times M} - \mathbf{W}\mathbf{W}^T\right)^{-1}$. It can be shown that the matrix $\boldsymbol{\Pi}$ is the orthogonal projection operator onto the column space of $\mathbf{U}^\nabla$. Let $\boldsymbol{\Pi} = \left(\boldsymbol{\Pi}^{(1)}, \ldots, \boldsymbol{\Pi}^{(M)}\right)^T$ and $\mathcal{R} = \left(\mathcal{R}^{(1)}, \ldots, \mathcal{R}^{(M)}\right)^T$, where $\boldsymbol{\Pi}^{(i)}$ with dimension $(L_i - 1) \times (L_{sum} - M)$ and $\mathcal{R}^{(i)}$ ($i = 1, \ldots, M$) with length $L_{sum} - M$, corresponds to the series $Y_{N_i}^{(i)}$. Then, Theorem B.6 (please see Appendix B), indicates that the linear projection $\mathscr{P}^{(v)} : \mathfrak{L}_r \mapsto \mathbb{R}^{L_{sum} - M}$ obtained by the following formula provides the continuation vectors for multivariate V forecasting.

$$\mathscr{P}^{(v)} Y = \begin{bmatrix} \boldsymbol{\Pi}^{(1)} Y_\Delta \\ \mathcal{R}^{(1)T} Y_\Delta \\ \vdots \\ \boldsymbol{\Pi}^{(M)} Y_\Delta \\ \mathcal{R}^{(M)T} Y_\Delta \end{bmatrix}, \quad Y \in \mathfrak{L}_r, \quad (2.20)$$

where $Y_\Delta^T = \left(Y_\Delta^{(1)}, \ldots, Y_\Delta^{(M)}\right)$, such that $Y_\Delta^{(i)}$ ($i = 1, \ldots, M$) denotes the last $L_i - 1$ entities of Y_i with length L_i. Using the above notation, the following algorithm produces the VMSSA-V forecasts.

VMSSA-V Forecasting Algorithm

1. Define the vectors Z_i as follows:

$$Z_i = \begin{cases} \widetilde{X}_i & \text{for } i = 1, \ldots, k \\ \mathscr{P}^{(v)} Z_{i-1} & \text{for } i = k+1, \ldots, k+h+L_{\max}-1, \end{cases} \quad (2.21)$$

where $L_{\max} = \max\{L_1, \ldots, L_M\}$.

2. Construct the matrix $\mathbf{Z} = [Z_1 : \ldots : Z_{K+h+L_{\max}-1}]$ and obtain its hankelization. Using this calculation we obtain $\hat{y}_1^{(i)}, \ldots, \hat{y}_{N+h+L_{\max}}^{(i)}$, $i = 1, \ldots, M$.
3. The elements $\hat{y}_{N_i+1}^{(i)}, \ldots, \hat{y}_{N_i+h}^{(i)}$ ($i = 1, \ldots, M$) form the h-step ahead VMSSA-V forecasts.

R Codes for VMSSA-V

The following R function, which must be defined after R functions VMSSA.Hankel(), V.SVD(), V.Group() and DiagAver(), can be used to obtain the VMSSA-V forecasts by R:

Program 2.14 Vector VMSSA forecasting R code

```
VMSSA.V<-function(k,Group,h,Y){
r<-length(Group)
Dec<-V.SVD(Y,k)
sigma<-Dec$d
d<-length(sigma[sigma>0])
U<-matrix(Dec$u,ncol=d)
V<-matrix(Dec$v,d,k)
hx<-V.Group(Y,k,Group)
if(is.list(Y)){
M<-length(Y);L<-NULL;N<-NULL
for(i in 1:M){
N[i]<-length(Y[[i]])
L[i]<-N[i]-k+1}}
else{
M<-ncol(Y);N<-rep(nrow(Y),M);L<-N-k+1}
L.last<-cumsum(c(0,L));L.first<-cumsum(c(0,(L-1)))
forecasts<-array(0,dim=c(h,M))
W<-array(U[cumsum(L),Group],dim=c(M,r))
Ud<-array(U[-cumsum(L),Group],dim=c((sum(L)-M),r))
```

```
R<-Ud%*%t(W)%*%solve(diag(M)-W%*%t(W))
pai<-Ud%*%t(Ud)+R%*%(diag(M)-W%*%t(W))%*%t(R)
Pv<-array(dim=c(sum(L),(sum(L)-M)))
for(j in 1:M){
Lj<-(L.first[j]+1):L.first[j+1]
Pv[(L.last[j]+1):L.last[j+1],]<-rbind(pai[Lj,],R[,j])}
G<-hx
for(l in (k+1):(max(N)+h))
G<-cbind(G,(Pv%*%G[-(L.last[1:M]+1),(l-1)]))
for(j in 1:M){
Lj<-(L.last[j]+1):L.last[j+1]
forecasts[,j]<-DiagAver(G[Lj,])[(N[j]+1):(N[j]+h)]}
forecasts
}
```

> **Example 2.7: Vector VMSSA forecasting for the motor data set**

The results of six new VMSSA-V forecasts for the real data of Example 2.3 by using $L = 9$ ($k = 10$) and first three eigentriples are:

```
round(VMSSA.V(10,c(1:3),6,motor),1)
       [,1]    [,2]    [,3]
[1,] 1325.4  1229.3  1664.7
[2,] 1443.7  1333.4  1834.8
[3,] 1568.2  1445.1  2015.7
[4,] 1699.5  1564.0  2208.1
[5,] 1838.1  1689.9  2412.5
[6,] 1984.4  1823.2  2629.4
```

2.5 Automated MSSA

As in the case of univariate SSA, it is possible to automate the selection of key parameters (see Sect. 1.5). The HMSSA-R and HMSSA-V automated forecasting algorithms are presented below, see Hassani and Mahmoudvand (2013).

2.5.1 MSSA Optimal Forecasting Algorithm

1. Consider M time series with identical series lengths of N, such that $Y_N^{(i)} = (y_1^{(i)}, \ldots, y_N^{(i)})$ ($i = 1, \ldots, M$).

2. For forecasting exercises split each time series into three parts leaving $\frac{2}{3}$rd for model training and testing, and $\frac{1}{3}$rd for validation.
3. Beginning with a fixed value of L, seek the combination of L and r which minimizes a loss function, $\mathscr{L}$, and thus represents the optimal MSSA choices for forecasting in a multivariate framework.
4. Finally, use the selected optimal L to decompose the series comprising of the validation set, and then select r singular values for reconstructing the less noisy time series, and use this newly reconstructed series for forecasting the remaining $\frac{1}{3}$rd of the data.

2.5.2 Automated MSSA R Code

We begin by presenting the automated HMSSA-R codes for finding the optimal L and r. The code shown here is tailored for three variate, $M = 3$, time series, however, it can be easily tweaked in order to incorporate more than three time series in one round. We have considered root mean square error as the loss function, but this can be easily replaced with any other loss function. We provide three R functions to find the optimal choices for HMSSA-R. The first function, RMSE(), produces the root mean squared error for HMSSA-R forecasts; the second and third functions, Opt.r() and Opt.choices(), are the main functions to produce the optimal choices.

Program 2.15 RMSE computation R code

```
RMSE<-function(Y,L,r,mse.size,h){
N<-nrow(Y);M<-ncol(Y)
forecasts<-NULL
train.size<-N-mse.size-h+1
test<-Y[(train.size+h):(train.size+mse.size+h-1),]
for(i in 1:mse.size){
train<-Y[1:(train.size+i-1),]
forec<-matrix(HMSSA.R(L,1:r,h,train),ncol=M)[h,]
forecasts<-rbind(forecasts,forec)
}
sqrt(sum(colMeans(as.matrix((test-forecasts)^2))))
}
```

Program 2.16 Optimal r searcher R code

```
Opt.r<-function(Y,L,mse.size,h){
opt.rmse<-array(0,dim=(L-1))
for(r in 1:(L-1))
opt.rmse[r]<-RMSE(Y,L,r,mse.size,h)
which.min(opt.rmse)
}
```

Program 2.17 Optimal MSSA choices finder R code

```
Opt.choices<-function(Y,L.vec,mse.size,h){
    wid<-length(L.vec)
    rmse<-array(0,dim=wid)
for(i in 1:wid){
r0<-Opt.r(Y,L.vec[i],mse.size,h)
rmse[i]<-RMSE(Y,L.vec[i],r0,mse.size,h)
}
window<-which.min(rmse)
r.opt<-Opt.r(Y,L.vec[window],mse.size,h)
L.opt<-L.vec[window]
list(optimal.r=r.opt,optimal.L=L.opt,root.mean.square.error=rmse)
}
```

Once the optimal L and r are determined, we can use those values with the R function HMSSA.R() to produce forecasts in the validation part.

> **Example 2.8: Optimal choices for Recurrent HMSSA forecasting for the motor data set**

Consider the first 18 observations from Example 2.3 to find the optimal choices for using HMSSA-R. Assuming 1-step ahead forecasting with 6 observations for obtaining the RMSE, the following R codes produce the optimal choices:

```
Opt.choices(motor,c(2:9),6,1)
$optimal.r
[1] 1

$optimal.L
[1] 2
```

```
$root.mean.square.error
[1] 166.4430 179.3497 188.9558 200.5602
[5] 177.1275 177.3263 181.0547 180.1823
```

Now you can use $L = 2$ and $r = 1$ to find HMSSA-R forecasts for observation 19. The result in R is:

```
HMSSA.R(2,1,1,motor)
[1] 1359.003 1322.152 1705.432
```

Note that in order to use an automated MSSA R code for other algorithms, it is enough to replace the lines including the forecasting method in R functions RMSE() Opt.r() and Opt.choices() with related forecasting method.

2.6 A Real Data Analysis with MSSA

In this section, we apply the R code and program of this chapter to analyse the US Air Passenger monthly data. The data are provided by the US Bureau of Transportation Statistics, and are obtainable via http://www.transtats.bts.gov/Data_Elements.aspx?Data=3. The data include domestic miles and international miles over the period Oct. 2002–Oct. 2015 ($N_1 = N_2 = 157$), and using R function scan() these can be entered in R. To illustrate, we used scan() to enter the data, with the last three observations shown below, where the data are in two vectors defined as domestic and international.

```
domestic<-scan()
 .
 .
 .
155: 57819058
156: 50016326
157: 54074201
158:
Read 157 items
> international<-scan()
 .
 .
 .
155: 27156489
```

```
156: 22404096


157: 21786384
158:
Read 157 items
```

Using the R function `ts()`, we define a bivariate time series, `airpass`, including these time series:

```
series<-cbind(domestic,international)
airpass<-ts(series,freq=12,start=c(2002,10))
```

Description of Time Series

Figure 2.1, which is produced by the following R code, shows the behaviour of the time series:

```
ts.plot(airpass,xlab="year",ylab="No. passengers",col=c(2,4))
legend("topleft",c("Domestic","International"),col=c(2,4),lty=1,bty="n")
```

Note that multivariate time series analysis is usually performed when one aims at modelling and forecasting two time series using the interactions and comovements between two series which obviously exists between domestically and internationally number of US passengers considered here. It is evident from the figure that there exists similar trend, in addition to the oscillatory components, in the structure of the both series.

Fig. 2.1 Monthly number of US passengers, domestically and internationally, sample Oct. 2002–Oct. 2015

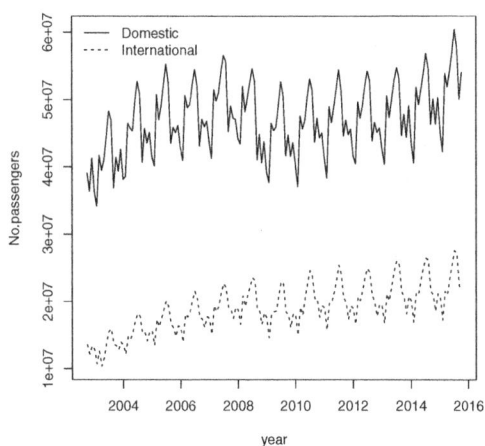

Fig. 2.2 Plot of Singular values of the trajectory matrix with Monthly number of US passengers, time series with $L = 72$

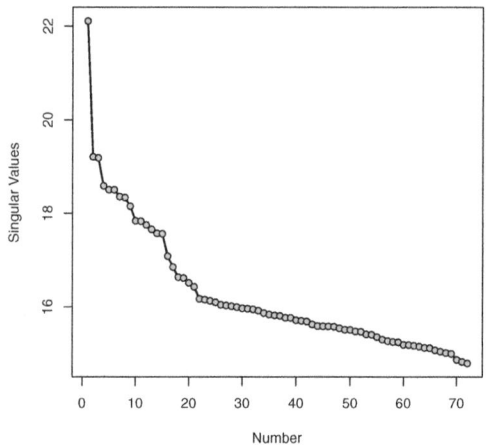

Finding the Optimal MSSA Choices

We first consider the scree plot of singular values with the trajectory matrix of the HMSSA algorithm. In order to obtain this plot, Programs 2.1 and 2.2 must be loaded together with Program 1.7 where H.SVD is used instead of SVD in the second line. The singular value plot is shown in Fig. 2.2, where $L = 72$.

A significant drop in values occurs around a candidate eigenvalue which could be interpreted as the start of the noise floor. To determine those eigenvalues related to the noise component, it is assumed that the eigenvalues are small and decline gradually. The behaviour of singular values in Fig. 2.2 encourages us to consider the first 21 singular values for filtering the series with the remainder as noise. Let us now check the optimal MSSA choices for forecasting the last 12 observations. In order to find this value for Recurrent HMSSA, load Programs 2.1–2.5, 2.11 and 2.15–2.17. Then it is sufficient to type the following R code:

```
Opt.choices(airpass,c(2:72),5,12)
```

Running this command takes about five minutes (depending on the speed of the computer in use) and produces the following results:

```
$optimal.r
[1] 11

$optimal.L
[1] 14

$root.mean.square.error
 [1] 2952004 4014611 5034832 5509866 5029147 7487224 2442005
 [8] 2073212 2042516 2571711 2592637 1902581 1898909 1974058
[15] 1970289 2128443 2480362 2465013 2692042 2582139 2300229
[22] 2309883 2086952 2102240 2398582 2688802 2618306 2659840
[29] 2614256 2622574 2584341 2626710 2871552 3042631 2790252
[36] 2427393 2286309 2956119 2900080 2964300 2919186 2793474
[43] 3052867 2912294 2952709 2771883 2882002 2904458 2970768
[50] 3096624 3088524 2974489 2904657 2934051 3136383 3241177
[57] 3409747 3450590 3632831 3808314 3664114 3727238 3510075
[64] 3535654 3712724 4120058 2978552 3168082 3217738 3216446
[71] 3211659
```

Let us now compare the performance of the choices: $L = 72, r = 21$ and $L = 14, r = 11$. Using Program 2.11, we find the forecast values of observations 146–157 for both selections. Computations were obtained with the following R codes:

```
forecast1<-HMSSA.R(Y=airpass[1:145,],L=72,1:21,h=12)
forecast2<-HMSSA.R(Y=airpass[1:145,],L=14,1:11,h=12)
```

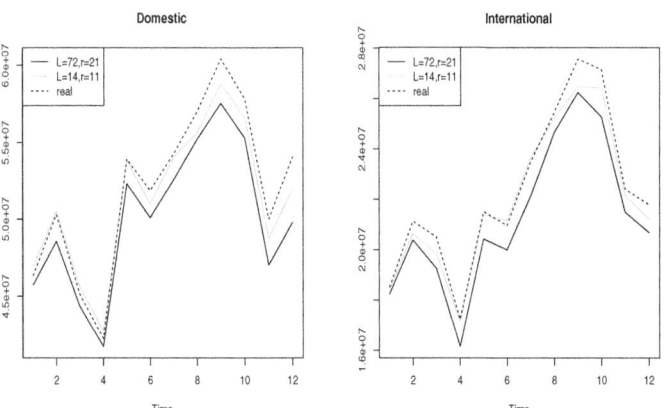

Fig. 2.3 Comparison of two forecasting scenarios with real observations

We are now able to compare the forecasts with the real observations and the results are shown in Fig. 2.3.

The left panel of this figure is obtained by the following R code:

```
ts.plot(ts(forecast1[,1]),forecast2[,1],airpass[146:157,1],col=c(1,8,1),lty=c(1,1,2))
legend("topleft",c("L=72,r=21","L=14,r=11","real"),col=c(1,8,1),lty=c(1,1,2))
title(main=("Domestic"))
```

and the right panel is produced by a similar R code as follows:

```
ts.plot(ts(forecast1[,2]),forecast2[,2],airpass[146:157,2],col=c(1,8,1),lty=c(1,1,2))
legend("topleft",c("L=72,r=21","L=14,r=11","real"),col=c(1,8,1),lty=c(1,1,2))
title(main=("International"))
```

CHAPTER 3

Applications of Singular Spectrum Analysis

Abstract This chapter presents three main applications of using Singular Spectrum Analysis (SSA): change point detection, gap filling/missing value imputation, and filtering/denoising. A concise description of the main idea along with technical background with various practical illustrations with associated R codes are given in this chapter. Both univariate and multivariate SSA R codes are provided for filtering and missing values imputation.

Keywords Change point detection · Gap filling · Missing value imputation · Denoising · Filtering

3.1 Introduction

In previous chapters, we considered forecasting by Singular Spectrum Analysis (SSA) whereas in this chapter, we explain how it can be used to monitor and detect change points in data generating process, and estimate missing data, usually referred to as gap filling. These are some examples of the flexibility of SSA. References for change point detection include Golyandina et al. (2001), Moskvina and Zhigljavsky (2003) and Mohammad and Nishida (2011); and references for gap filling include Mahmoudvand and Rodrigues (2016b), Kondrashov and Ghil (2006) and Golyandina and Osipov (2007). We also explain how SSA can be used for denoising, references for which include Alonso et al. (2005), Hassani et al. (2010), Mahmoudvand and Rodrigues (2016a) and Kume and Nose-Togawa (2014).

3.2 Change Point Detection

We start with a description of the structural changes in time series that we are attempting to detect. Assuming that a time series is governed by a linear recurrent formula (LRF), what would happen if suddenly this time series stops following the original LRF? As a result, the series as a whole stops being homogeneous and the problem of studying this heterogeneity arises: this is the change-point detection problem.

3.2.1 A Simple Change Point Detection Algorithm

As in SSA, we assume a time series given by $Y_N = (y_1, \ldots, y_N)$, but now with a single break point or multiple break points at unknown times $b_1, b_2, \ldots \in (1, \ldots, N)$, so that $Y_N = (y_1, \ldots, y_{b_1}, y_{b_1+1}, \ldots, y_{b_2}, y_{b_2+1}, \ldots, y_N)$. According to the basic algorithm that was proposed in Golyandina et al. (2001) and Moskvina and Zhigljavsky (2003), two subseries of Y_N will be selected and the column spaces of their trajectory matrices will be compared. If they are far from each other we conclude that change has occurred in the structure of the series. The way of choosing these subseries, the comparison measure and choosing some corresponding parameters, are the most important requisites of change point detection by SSA. The following algorithm is a simple way to use SSA for change-point detection in a univariate time series, but interested readers can refer to Golyandina et al. (2001), or Moskvina and Zhigljavsky (2003) to understand more details and theory of this application of SSA.

(i) Consider a time series $Y_N = (y_1, \ldots, y_N)$;
(ii) Define subseries sizes B and T, where $B, T < N$;
(iii) Set $t = 0$ and f a positive integer such that $f \leq B - L$;
(iv) Using L, where $L \leq B/2$, construct two trajectory matrices $\mathbf{X}_{Base}$ and $\mathbf{X}_{Test}$ by considering subseries $(y_{t+1}, \ldots, y_{t+B})$ and $(y_{t+f+1}, \ldots, y_{t+f+T})$, respectively;
(v) Perform SVD on the matrix $\mathbf{X}_{Base}$, and construct the matrix $\mathbf{U}$ by stacking the first r left eigenvectors;

(vi) Compute the statistic:

$$D_t = 1 - \frac{\sum_{j=1}^{K_2} Z'_j Z_j}{\sum_{j=1}^{K_2} X'_{Test,j} X_{Test,j}}, \qquad (3.1)$$

where $K_2 = T - L + 1$ and $Z_j = \mathbf{U}' X_{Test,j}$ where $X_{Test,j}$ is the j-th column of the matrix $\mathbf{X}_{Test}$;

(vii) Set $t = t + 1$ and repeat steps (iii) to (vi);

(viii) Analyse D_t: the larger D_t indicates the change point in the associated time intervals.

Constraint on Parameters in Change-Point Algorithm

There are several parameters that must be determined for detecting the change points by SSA. Here, we consider the following conditions:

B: the size of the base subseries depends on the nature of the structural changes we are looking for. As a general rule, a smaller value provides gradual changes whereas a large value smooths out smaller changes;

L: window length, it can be smaller than or equal to $B/2$;

f: first observation number for the test subseries, which can be larger than $B - L + 1$;

T: size of the test subseries, which can be bigger than max $\{f, B - f\}$;

r: grouping parameter, must be less than L.

3.2.2 Change-Point Detection R Code

Program 3.1 Implements the change point algorithm to obtain the D_t statistic

Program 3.1 Change point detection R code

```
UniHankel<-function(Y,L){
k<-length(Y)-L+1
outer((1:L),(1:k),function(x,y) Y[(x+y-1)])
}

D.Comp<-function(L,r,Y,B,f,T,t){
N<-length(Y)
X.Base<-UniHankel(Y[(t+1):(t+B)],L)
X.Test<-UniHankel(Y[(t+f+1):(t+f+T)],L)
```

```
K<-T-L+1
D0<-D1<-0
U<-matrix(svd(X.Base)$u,L,L)
U.r<-U[,1:r]
for(j in 1:K){
D0<-D0+t(X.Test[,j])%*%X.Test[,j]
D1<-D1+t(X.Test[,j])%*%U.r%*%t(U.r)%*%X.Test[,j]
}
1-D1/D0
}


Possible.D<-function(L,r,Y,B,f,T){
D<-0
N<-length(Y)
for(t in 0:(N-T-f))
D<-c(D,D.Comp(L,r,Y,B,f,T,t))
D
}
```

Example 3.1: Change point detection in an artificial data set

In this example, we construct a series of size 326 and depict its plot by the following R codes:

```
sim<-c(1:200,.5*c(399:300),2*c(75:100))
plot(ts(sim),xlab="Obs. No", ylab="Artificial data")
```

The data are plotted in the left panel of Fig. 3.1 which suggests two change points located at observations 200 and 300. Below, we first copy the R Program 3.1 into the R console and then use it with parameters $L = 35$, $r = 14$, $B = 70$, $T = 50$ and $f = 40$ to find and plot the possible change points as below:

```
D<-Possible.D(L=30,r=14,Y=sim,B=70,f=40,T=50)
plot(ts(D,start=71),xlab="Obs. No", ylab=expression(D[t]))
```

Figure 3.1 confirms that the detected change points coincide with the real change points in the data.

3 APPLICATIONS OF SINGULAR SPECTRUM ANALYSIS   91

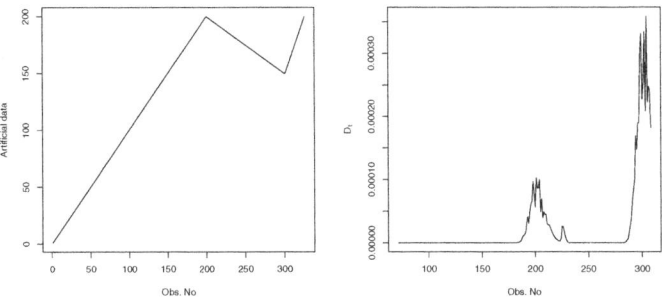

Fig. 3.1 Initial data (left side) and change-point detection statistic D_t (right side) in Example 3.1

Example 3.2: Plotting D statistics in a real data set

Consider again the monthly UK government security yield from Jan. 1985 to Dec. 2015, UKYield, as in Sect. 1.8.2. The length of the time series is 372 and we aim to assess this for the presence of structural changes in this series. To illustrate the procedure, we fit the trend by the first eigentriple, with for example $L = 72$, using the following R code:

```
Approx<-SSA.Rec(UKYield,72,c(1))$Approximation
Data<-cbind(UKYield,Approx)
Series<-ts(Data,frequency=12,start=c(1985,1))
plot.ts(Series[,1],type="l",xlab= "Time", ylab= "UK Yield")
legend("bottomleft",horiz=FALSE,lwd=c(1,3),c("Initial","EF 1"))
lines(Series[,2],lwd=2)
```

The left panel of the Fig. 3.2 shows the results. Below, we first copy the R Program 3.1 into the R console and then use it with parameters $L = 72$, $r = 13$, $B = 144$, $T = 90$ and $f = 73$ to find the possible change points as follows:

```
D<-Possible.D(L=72,r=13,Y=UKYield,B=144,f=73,T=90)
plot(ts(D,start=c(1997,1),freq=12),ylab=expression(D[t]),xlab="Year")
```

This code produces the right side plot in Fig. 3.2, which suggests three time periods where changes in trend may have occurred, corresponding to the years 2008, 2011 and 2013–2014. It should be mentioned that D_t is considered to be equivalent to a change by point $73 + t$.

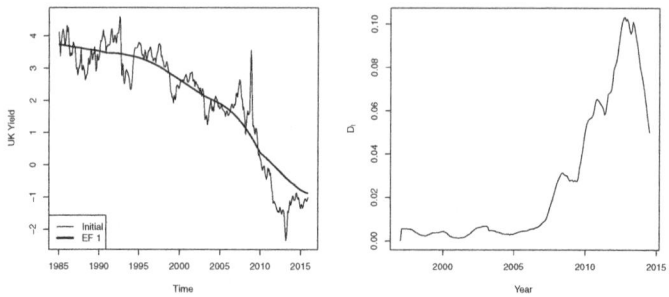

Fig. 3.2 Fitting the trend (left side) and the change-point detection statistic D_t (right side) in Example 3.2

3.3 Gap Filling with SSA

In this section, we consider the application of SSA to impute missing values in time series; see, for example Schoellhamer (2001), Kondrashov and Ghil (2006), Golyandina and Osipov (2007), Rodrigues and de Carvalho (2013), and Shen et al. (2015). Here, we outline two ways of gap filling:

1. A reconstruction based technique in which missing values are replaced with zero at the first step and then updated using basic SSA filtering and this cycle is repeated until convergence is reached.
2. A forecasting technique in which the missing data are replaced with data based on forecasting with the remaining data.

Examples 3.3 and 3.4, which follow, illustrate the use of these methods.

> **Example 3.3: Missing imputation in a real data set**

The first data set we consider is the annual Research and Development, R&D, expenditure (% GDP) for Denmark for the period 1996–2013. The data are provided by Eurostat (http://ec.europa.eu/eurostat/data/ database). The data were entered in R and saved as the object RD, which is shown below:

```
RD
 [1]  1.79 1.81 1.89 2.01 2.13 2.19 2.32 2.44 2.51 2.42 2.39
[12]  2.40 2.51 2.78 3.07 2.94 2.97 3.00 3.06
```

Assume that the fifth observation is missing. In order to apply the reconstruction based technique, the following R code can be used. The singular values of the trajectory matrix for this data set, when the fifth observation is replaced with zero are given by the second line in the code, Sing.plt() function which is explained in Chapter 1, Sect. 1.3.2. This command produces Fig. 3.3 which suggests to use the first five eigentriples to reconstruct the series.

```
RD[5]<-0
Sing.plt(RD,L=9)
estimate<-0
for(j in 1:50){
RD[5]<-SSA.Rec(RD,L=9,1:5)$Approximation[5]
estimate[j]<-RD[5]}
estimate
```

The results of this R code are given below, with the estimation procedure converging after 60 iterations to provide the value 2.0819782.

```
[1]  0.1536593 0.2964765 0.4296572 0.5543407 0.6716236
[6]  0.7825804 0.8882744 0.9897468 1.0879597 1.1836557
[11] 1.2771104 1.3678338 1.4544375 1.5349290 1.6073947
[16] 1.6706519 1.7245455 1.7700614 1.8119589 1.8733102
[21] 1.9391700 2.0009073 2.0396938 2.0601980 2.0707377
[26] 2.0761592 2.0789594 2.0804101 2.0811631 2.0815544
[31] 2.0817578 2.0818635 2.0819185 2.0819472 2.0819620
```

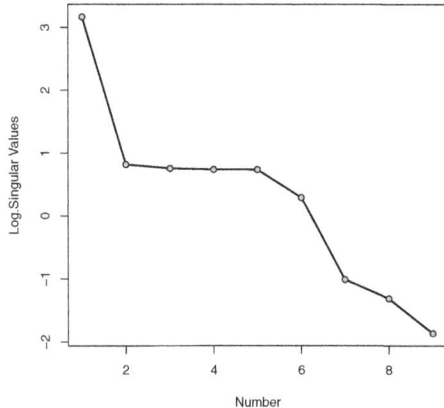

Fig. 3.3 Logarithm of the singular values of the trajectory matrix in Example 3.3, when $L = 9$

```
[36]  2.0819698  2.0819738  2.0819759  2.0819770  2.0819776
[41]  2.0819778  2.0819780  2.0819781  2.0819781  2.0819781
[46]  2.0819782  2.0819782  2.0819782  2.0819782  2.0819782
```

Example 3.4: Missing imputation in a real data set

In this example, we consider the real interest rate in Japan for the period 1961–2013, $N = 54$, in which assume the observations 27 and 28 are missing. (The data are provided by The World Bank, see http://www.data.worldbank.org/country/japan?display=default.) We entered this data in R under the object name `IntRt.Jp` with the content of this object given below:

```
IntRt.Jp
 [1]   0.1957115   3.8500691   2.1573192   2.4385193    2.5346101
 [6]   2.3886445   1.7241132   2.4193014   2.8463289  -12.2299202
[11]   2.3805302   1.3644426  -4.8969671  -9.6819073    1.7927210
[16]   0.2298723   0.7611867   1.7361470   3.5198490    2.7574023
[21]   4.5224616   5.6675890   6.1575859   4.9193017    5.5394905
[26]   4.1693491   5.3205652   4.6863158   2.9940818    4.4978776
[31]   4.7994542   4.4912472   4.4077604   4.0137587    4.2620611
[36]   3.2324837   1.8428085   2.3766787   3.4788431    3.3566422
[41]   3.2056841   3.4689681   3.5985456   3.1618116    2.9651673
[46]   2.8178660   2.8404274   3.2153825   2.2347971    3.8459984
[51]   3.4153972   2.3606647   1.8623912 -10.7290722
```

The data are plotted by using the following R command and shown in Fig. 3.4.

```
plot.ts(ts(IntRt.Jp,start=1961,freq=1),ylab="Real Interest Rate")
```

In this example, we use MSSA to impute the missing data in the series. To implement MSSA define two series: (i) first, the series including observations 1–26, and (ii) second, the series including observations 29–54, but in a reverse order. The MSSA forecasting procedure is then used to obtain the 2-step ahead forecast, which can be converted as estimates of the missing values by averaging forecasts. The results are presented alternatively with both Recurrent and Vector HMSSA. We start the procedure by the number of eigentriples that are required for reconstruction. The following R code provides the singular values of the trajectory matrix:

```
Rate1<-IntRt.Jp[1:26]
Rate2<-IntRt.Jp[54:29]
MSing.plt<-function(Y,L){
```

3 APPLICATIONS OF SINGULAR SPECTRUM ANALYSIS

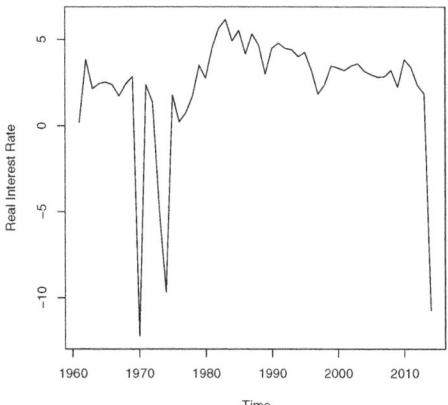

Fig. 3.4 Real interest rate in Japan during 1961–2014

```
lambda<-log(H.SVD(Y,L)$d)
d<-length(lambda)
win<-1:d
plot.new()
plot.window(xlim=range(win),ylim=range(lambda))
usr=par("usr")
rect(usr[1],usr[3],usr[2],usr[4])
lines(win,lambda,lwd=2)
points(win,lambda,pch=21,bg="gray")
axis(1)
axis(2)
box()
title(xlab="Number")
title(ylab="Log. Singular Values")}
MSing.plt(list(Rate1,Rate2),L=13)
```

The results are shown in Fig. 3.5

Figure 3.5 suggests using the first four eigentriples to operate HMSSA on the series. The following results are obtained using the R functions HMSSA.R() and HMSSA.V:

```
HMSSA.R(L=13,Group=1:4,h=2,Y=list(pat1,pat2))
     [,1]      [,2]
[1,] 6.676902  3.464212
[2,] 5.917000  3.249537
> HMSSA.V(L=13,Group=1:4,h=2,Y=list(pat1,pat2))
     [,1]      [,2]
[1,] 6.931000  3.636430
[2,] 5.723584  3.247136
```

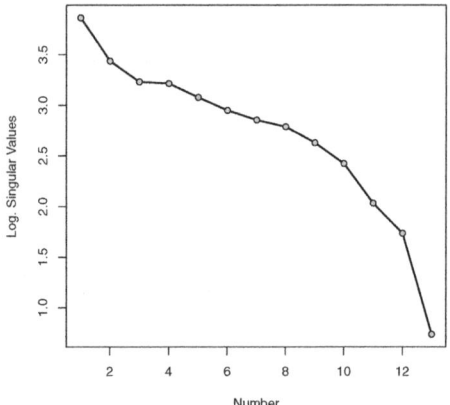

Fig. 3.5 Scree plot of singular values of the HMSSA trajectory matrix, when $L = 13$ in Example 3.4

Denote by $\hat{y}_{27}$ and $\hat{y}_{28}$ the estimates of missing data, HMSSA.R results in:

$$\hat{y}_{27} = \frac{6.676902 + 3.249537}{2} = 4.96322 \ , \ \hat{y}_{28} = \frac{5.917000 + 3.464212}{2} = 4.690606$$

and HMSSA.V results in:

$$\hat{y}_{27} = \frac{6.931 + 3.249537}{2} = 5.089068 \ , \ \hat{y}_{28} = \frac{5.723584 + 3.636430}{2} = 4.680007$$

Note that the real values of the above estimated missing values are $y_{27} = 5.3205$ and $y_{28} = 4.6863$. These outputs show that the estimates based on Recurrent HMSSA and Vector HMSSA approach are close together.

3.4 Denoising by SSA

Inferential statistics includes techniques to measure the relationships between particular variables. For example, regression analysis may be used to model whether a change in advertising explains the variation in sales. It is clear that the existence of a few outliers or a significant noise level may reduce the performance of the statistical methods. In order to deal with this problem, we can use filtering or weighting data. If the noise has been significantly reduced in the noisy data, then this approach is expected to give much better results than other approaches. The problem of filtering data to obtain noiseless data with minimum loss of information has been widely studied (see, e.g. Rendall et al. 2014 and references therein). Recent research has shown that SSA can be used as an alternative to traditional digital filtering methods (see, e.g. Alonso et al. 2005; Kume and Nose-Togawa 2014). Let us see the application of SSA in correlation analysis.

3.4.1 Filter Based Correlation Coefficients

Let $(x_1, y_1), \ldots (x_n, y_n)$ be n observations from a bivariate distribution. Assume that:

$$\begin{cases} x_t = s_x(t) + \epsilon_x(t) \\ y_t = s_y(t) + \epsilon_y(t), \end{cases} , t = 1, \ldots, n, \quad (3.2)$$

where $s_x(t)$ and $s_y(t)$ are the signal components and $\epsilon_x(t)$ and $\epsilon_y(t)$ are the noise terms. Then we define filter based correlation coefficient as the correlation between $s_x(t)$ and $s_y(t)$. Note that we expect to obtain a reliable correlation close to the true value, if we are able to remove the noise terms from the variables. Denote by $(\hat{s}_x(t), \hat{s}_y(t))$, for $t = 1, \ldots, n$, the observations filtered by SSA, we then use Eq. (4.2) to compute the filter based correlation coefficient:

$$R_{x,y} = \frac{\sum_{t=1}^{n} (\hat{s}_x(t) - \bar{\hat{s}}_x)(\hat{s}_y(t) - \bar{\hat{s}}_y)}{\sqrt{\sum_{t=1}^{n} (\hat{s}_x(t) - \bar{\hat{s}}_x)^2 \sum_{t=1}^{n} (\hat{s}_y(t) - \bar{\hat{s}}_y)^2}}. \quad (3.3)$$

Using Programs 1.1–1.5 and Programs 2.1–2.10, we can use the following R function to obtain the SSA-based linear correlation between two vectors.

Program 3.2 SSA-Based correlation R code

```
SSA.cor<-function(x,y,L.u,ru.x,ru.y,L.h,rh,k.v,rv){
x.ssa<-SSA.Rec(x,L.u,1:ru.x)$Approximation
y.ssa<-SSA.Rec(y,L.u,1:ru.y)$Approximation
xy.hmssa<-HMSSA.Rec(cbind(x,y),L.h,1:rh)$Approximation
xy.vmssa<-VMSSA.Rec(cbind(x,y),k.v,1:rv)$Approximation
ssa.cor<-cor(x.ssa,y.ssa)
hmssa.cor<-cor(xy.hmssa[,1],xy.hmssa[,2])
vmssa.cor<-cor(xy.vmssa[,1],xy.vmssa[,2])
```

```
pearson<-cor(x,y)
data.frame(ssa.cor,hmssa.cor,vmssa.cor,pearson)
}
```

> **Example 3.5: Correlation analysis based on simulated data**

In this example, we use the following model to simulate a series of data:

$$s_x(t) = e^{\frac{t}{25}} \sin(2\pi t/5) \, , \ t = 1,\ldots,50, s_y(t) = 1 + 2s_x(t)$$

In the context of the system Eq. (3.2), we generate the error terms from Gaussian noise with mean zero and $\sigma_z^2 = var(s_z)/SNR_z$, where SNR_z is the signal to noise ratio, and $var(s_z) = \sum_{t=1}^{50}(s_z(t) - \bar{s}_z)^2/49$. The following R code produces the data set.

```
sig.x<-exp((1:50)/25)*sin(2*pi*(1:50)/5)
sig.y<-2*sig.x+1
```

Then plot this data by the R code:

```
plot(sig.x,sig.y)
```

which produces the left normal plot in Fig. 3.6. Assuming $SNR_x = 2$ and $SNR_y = 4$, add noise to this data and plot noisy data as below:

```
x<-sig.x+rnorm(length(sig.x),0,sd(sig.x)/sqrt(2))
y<-sig.y+rnorm(length(sig.x),0,sd(sig.y)/2)
plot(x,y)
```

The right normal plot in Fig. 3.6 shows the noisy data set.

In order to determine correct grouping in this example, we can use SVD and find the number of positive singular values. This is achieved by the following R code:

```
round(svd(UniHankel(sig.x,25))$d,3)
 [1] 42.326 41.185  0.000  0.000  0.000  0.000  0.000  0.000
 [9]  0.000  0.000  0.000  0.000  0.000  0.000  0.000  0.000
[17]  0.000  0.000  0.000  0.000  0.000  0.000  0.000  0.000
[25]  0.000

round(svd(HMSSA.Hankel(cbind(sig.x,sig.y),33))$d,3)
 [1] 90.953 89.621 24.162  0.000  0.000  0.000  0.000  0.000
 [9]  0.000  0.000  0.000  0.000  0.000  0.000  0.000  0.000
```

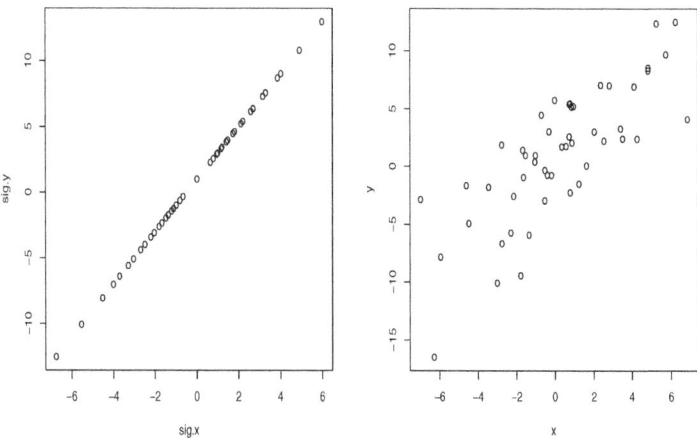

Fig. 3.6 Real signal (right side) and noisy data (left side) in Example 3.5

```
[17]  0.000  0.000  0.000  0.000  0.000  0.000  0.000  0.000
[25]  0.000  0.000  0.000  0.000  0.000  0.000  0.000  0.000
[33]  0.000

round(svd(VMSSA.Hankel(cbind(sig.x,sig.y),35))$d,3)
[1]  90.025  86.599  23.572  0.000  0.000  0.000  0.000  0.000
[9]   0.000   0.000   0.000  0.000  0.000  0.000  0.000  0.000
[17]  0.000   0.000   0.000  0.000  0.000  0.000  0.000  0.000
[25]  0.000   0.000   0.000  0.000  0.000  0.000  0.000  0.000
```

The results show that the number of singular values for reconstruction of $s_x(t)$ and $s_y(t)$ by univariate SSA are 2 and 3, respectively. In addition, both HMSSA and VMSSA reconstruct the signal by 3 eigentriples. Now, considering the window lengths $L = 25, 33$ and 16 for univariate SSA, HMSSA and VMSSA, respectively, the following R code, Program 3.2, will produce SSA-based correlations and the Pearson correlation in the simulated noisy data set in the Example 3.5. The results show that using both HMSSA and VMSSA, we obtain more accurate conclusion and the noise level for calculating the correlation between $s_x(t)$ and $s_y(t)$ is less effective.

```
SSA.cor(x,y,L.u=25,ru.x=2,ru.y=3,L.h=33,rh=3,k.v=35,rv=3)

  ssa.cor hmssa.cor vmssa.cor  pearson
1 0.9396713 0.9923791 0.9941279 0.787572
```

> **Example 3.6: Correlation analysis in a real data set**

In this example, we consider the relationship between the number of technicians in R&D per million of the population of Japan and high technology exports in Japan (in terms of the % of manufactured exports). These data are available via World Bank and period of study is 1996–2013. We entered data in R and named by `tech` and `export`. The data are shown below:

```
tech
 [1] 672.3516 668.0532 693.0988 673.6236 628.0174 545.7780
 [7] 531.0120 532.6188 576.6682 564.8661 581.0602 589.5402
[13] 593.1609 587.4390 587.9447 564.6080 517.7607 519.2189

exports
 [1] 26.14833 26.40513 26.15223 26.64937 28.68867 26.59550
 [7] 24.77556 24.42805 24.10432 22.98115 22.05696 18.40781
[13] 17.31434 18.75847 17.97278 17.45941 17.40937 16.78429
```

Figure 3.7 shows the relationship between exports and number of technicians in Japan, as produced by the following R code:

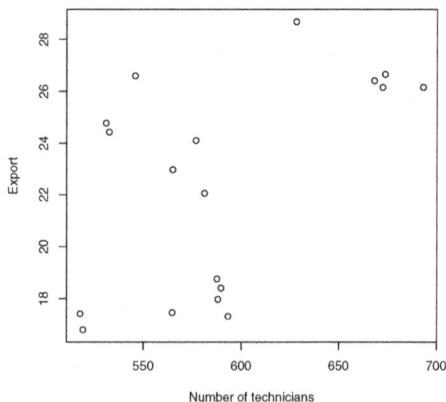

Fig. 3.7 Scatterplot for number of technicians and export percent in Example 3.6

```
plot(patent,export,xlab="No. patent application", ylab="Export")
```

The results of the SSA-based correlations using Program 3.2 are produced below:

```
SSA.cor(export,tech,L.u=9,ru.x=1,ru.y=1,L.h=12,rh=1,k.v=13,rv=1)
    ssa.cor hmssa.cor vmssa.cor    pearson
1 0.9545738 0.9921003 0.9966654 0.5181722
```

It is widely accepted that noise reduction matters in measuring dependencies between two time series. The above examples, using both simulated and real data, confirmed the significant effect of noise reduction on the linear and nonlinear measure of dependencies. Thes examples illustrate that measuring correlation using a multivariate system and filtered series is more effective.

CHAPTER 4

More on Filtering and Forecasting by SSA

Abstract The core intention of this chapter is providing the reader with more theoretical information regarding technical background of the Singular Spectrum Analysis (SSA) approach so that each part of the presented technique can be easily extended/modified according to the users' desirer/applications. The possibility of using two window lengths for reconstruction and forecasting stages of SSA and outlier detection are presented as two illustrations. More information on theoretical aspect of SSA is also presented in Appendix B.

Keywords Outlier detection · Technical aspect of SSA · Forecasting formula · Filtering formula

4.1 Introduction

Although the Singular Spectrum Analysis (SSA) algorithms seem to be simple, but understanding of what they do and how they fit among the other time series analysis techniques is by no means simple. We have spent 10 years working on this topic. As a whole, this book is a broad overview of what we have learned about SSA. Our experiences in using SSA for analysing different time series have convinced us SSA can be extremely useful.

Since the advent of modern data analytic methods, researchers expect to see a variable importance ranking for all models. In this chapter, we show how we can obtain the coefficients for filtering and forecasting.

4.2 Filtering Coefficients

In Chapter 1, we used SSA to construct a filtered out time series. Let us see a new representation for filtering time series by first r eigentriples. Consider the eigenvectors of $\mathbf{X}$, as $U_j = [u_{1,j}, \ldots, u_{L,j}]'$ for $j = 1, \ldots, d$, and

$$\mathbf{X}_j = U_j U_j' \mathbf{X} = \begin{bmatrix} u_{1,j} \sum_{t=1}^{L} u_{t,j} y_t & u_{1,j} \sum_{t=1}^{L} u_{t,j} y_{t+1} & \cdots & u_{1,j} \sum_{t=1}^{L} u_{t,j} y_{t+k-1} \\ u_{2,j} \sum_{t=1}^{L} u_{t,j} y_t & u_{2,j} \sum_{t=1}^{L} u_{t,j} y_{t+1} & \cdots & u_{2,j} \sum_{t=1}^{L} u_{t,j} y_{t+k-1} \\ \vdots & \vdots & \cdots & \vdots \\ u_{L,j} \sum_{t=1}^{L} u_{t,j} y_t & u_{L,j} \sum_{t=1}^{L} u_{t,j} y_{t+1} & \cdots & u_{L,j} \sum_{t=1}^{L} u_{t,j} y_{t+k-1} \end{bmatrix}. \quad (4.1)$$

Then, using Eq. (1.3) and performing diagonal averaging on the matrix $\sum_{j=1}^{r} \mathbf{X}_j$ we can write filtered out time series $\tilde{y}_t$ as below

$$\tilde{y}_t = \frac{1}{s_2 - s_1 + 1} \sum_{p=s_1}^{s_2} \sum_{i=1}^{L} \sum_{j=1}^{r} u_{i,j} u_{p,j} y_{t+i-p}, \quad t = 1, \ldots, N. \quad (4.2)$$

Equation (4.2) shows that we can write smoothed series as a linear combination of the original series (see also Harris and Yan 2010; Golyandina and Zhigljavsky 2013). Denoting by c_i the weight of y_i in the smoothed version of the series, we can rewrite Eq. (4.2) as

$$\tilde{y}_t = \begin{cases} \sum_{i=1}^{t+L-1} c_i y_i, & t = 1, \ldots, L \\ \sum_{i=t-L+1}^{t+L-1} c_i y_i, & t = L+1, \ldots, k \\ \sum_{i=t-L+1}^{N} c_i y_i, & t = k+1, \ldots, N. \end{cases} \quad (4.3)$$

Equation (4.3) gives the number of observations that are used to find the smoothed values of y_t, i.e, $\tilde{y}_t$:

$$\begin{cases} L+t-1 & t = 1, \ldots, L \\ 2L-1, & t = L+1, \ldots, k \\ N-t+L, & t = k+1, \ldots, N. \end{cases} \quad (4.4)$$

4 MORE ON FILTERING AND FORECASTING BY SSA   105

So, besides using all observations to compute the weights c_i, we consider at least L observations to reconstruct each value y_i by SSA. Therefore, using the law of large numbers, the reconstruction by SSA produces better approximation of the signal when L is as large as possible.

Below we provided a program R code function Fw to obtain the weights of observations for filtering by SSA. The output of this function is a matrix $N \times N$ in which the rows of this matrix show the weights of observations for filtering.

Program 4.1 Filtering weights R code

```
Fw<-function(Y,L,r){
N<-length(Y)
X<-UniHankel(Y,L)
U<-matrix(svd(X)$u[,1:r],L,r)
C<-matrix(0,N,N)
for(t in 1:N){
s1<-max(1,(t-N+L))
s2<-min(L,t)
for(p in s1:s2)
C[t,(t+1-p):(t+L-p)]<-C[t,(t+1-p):(t+L-p)]+U%*%U[p,]
C[t,]<-C[t,]/(s2-s1+1)
}
C
}
```

> **Example 4.1: Filtering weights**

Define time series Y as below which shows Turnover-RENAULT during 1985–1999 (data are per thousands).

```
Y<-c(253.2,313.1,322.7,296.2,370.3,443.8,426.8,453.6,553.4,
+ 588.6, 646.8, 850.0,1106.7,1184.5,1425.1)
```

Then, using $L = 7$ and $r = 2$ and with R code presented in Program (4.1), which must be executed after running Programs (1.1) we have:

```
> round(Fw(Y,7,2),2)
```

	[,1]	[,2]	[,3]	[,4]	[,5]	[,6]	[,7]	[,8]	[,9]	[,10]
[1,]	0.25	0.27	0.24	0.20	0.09	-0.02	-0.10	0.00	0.00	0.00
[2,]	0.13	0.27	0.26	0.23	0.15	0.04	-0.06	-0.05	0.00	0.00
[3,]	0.08	0.18	0.26	0.24	0.19	0.11	0.01	-0.04	-0.03	0.00
[4,]	0.05	0.11	0.18	0.24	0.21	0.16	0.09	0.01	-0.03	-0.03
[5,]	0.02	0.06	0.11	0.17	0.22	0.21	0.17	0.07	0.01	-0.02
[6,]	0.00	0.01	0.05	0.11	0.17	0.24	0.24	0.15	0.06	0.00
[7,]	-0.01	-0.02	0.00	0.05	0.12	0.21	0.29	0.21	0.12	0.05
[8,]	0.00	-0.01	-0.02	0.00	0.05	0.12	0.21	0.29	0.21	0.12
[9,]	0.00	0.00	-0.01	-0.02	0.00	0.05	0.12	0.21	0.29	0.21
[10,]	0.00	0.00	0.00	-0.02	-0.02	0.00	0.06	0.15	0.24	0.29
[11,]	0.00	0.00	0.00	0.00	-0.02	-0.02	0.01	0.07	0.17	0.24
[12,]	0.00	0.00	0.00	0.00	0.00	-0.03	-0.03	0.01	0.09	0.16
[13,]	0.00	0.00	0.00	0.00	0.00	0.00	-0.03	-0.04	0.01	0.05
[14,]	0.00	0.00	0.00	0.00	0.00	0.00	0.00	-0.05	-0.06	-0.03
[15,]	0.00	0.00	0.00	0.00	0.00	0.00	0.00	0.00	-0.10	-0.10

	[,11]	[,12]	[,13]	[,14]	[,15]
[1,]	0.00	0.00	0.00	0.00	0.00
[2,]	0.00	0.00	0.00	0.00	0.00
[3,]	0.00	0.00	0.00	0.00	0.00
[4,]	0.00	0.00	0.00	0.00	0.00
[5,]	-0.02	0.00	0.00	0.00	0.00
[6,]	-0.02	-0.02	0.00	0.00	0.00
[7,]	0.00	-0.02	-0.01	0.00	0.00
[8,]	0.05	0.00	-0.02	-0.01	0.00
[9,]	0.12	0.05	0.00	-0.02	-0.01
[10,]	0.20	0.10	0.03	-0.01	-0.02
[11,]	0.29	0.19	0.08	0.01	-0.01
[12,]	0.23	0.31	0.18	0.08	0.01
[13,]	0.14	0.24	0.35	0.20	0.08
[14,]	0.03	0.15	0.30	0.44	0.20
[15,]	-0.05	0.04	0.23	0.41	0.58

Note that the weights of y_t for finding filtered out $\tilde{y}_t$ is $\frac{r}{L}$ when $t = L, L+1, \ldots, N-L+1$. Because we have $s_1 = \max\{1, t-N+L\} = 1$ and $s_2 = \min\{t, L\} = L$ when $t = L, L+1, \ldots, N-L+1$. Using Eq. (4.2) the weight for y_t is:

$$\frac{1}{L}\sum_{i=1}^{L}\sum_{j=1}^{r} u_{i,j}^2 = \frac{1}{L}\sum_{j=1}^{r}\sum_{i=1}^{L} u_{i,j}^2 = \frac{1}{L}\sum_{j=1}^{r} U_j^T U_j = \frac{r}{L}.$$

4.3 Forecast Equation

Recall that Recurrent and Vector SSA forecasts could be found by Eqs. (1.11) and (1.15), respectively. Let us now find another useful recursive formula for Recurrent and Vector SSA forecasts.

4.3.1 Recurrent SSA Forecast Equation

By denoting $\widetilde{\mathbf{X}}$ the trajectory matrix of the time series $\{\tilde{y}_1, \ldots, \tilde{y}_N, \hat{y}_{N+1}, \ldots, \hat{y}_{N+h}\}$, including N reconstructed values and h out-of-sample predicted values, we have

$$\widetilde{\mathbf{X}} = \begin{bmatrix} \tilde{y}_1 & \cdots & \tilde{y}_K & \tilde{y}_{K+1} & \tilde{y}_{K+2} & \cdots & \hat{y}_{N+h-L+1} \\ \vdots & \cdots & \vdots & \vdots & \vdots & \cdots & \vdots \\ \tilde{y}_{L-1} & \cdots & \tilde{y}_{N-1} & \tilde{y}_N & \hat{y}_{N+1} & \cdots & \hat{y}_{N+h-1} \\ \tilde{y}_L & \cdots & \tilde{y}_N & \hat{y}_{N+1} & \hat{y}_{N+2} & \cdots & \hat{y}_{N+h} \end{bmatrix}. \qquad (4.5)$$

Now note that the $(K+1)^{\text{th}}$ column of matrix (4.5) can be obtained by

$$\begin{bmatrix} \tilde{y}_{K+1} \\ \vdots \\ \tilde{y}_N \\ \hat{y}_{N+1} \end{bmatrix} = \begin{bmatrix} \mathbf{0} & \mathbf{I} \\ 0 & A^T \end{bmatrix} \begin{bmatrix} \tilde{y}_K \\ \vdots \\ \tilde{y}_{N-1} \\ \tilde{y}_N \end{bmatrix}, \qquad (4.6)$$

and the $(K+2)^{\text{th}}$ column of matrix (4.5) by

$$\begin{bmatrix} \tilde{y}_{K+2} \\ \vdots \\ \hat{y}_{N+1} \\ \hat{y}_{N+2} \end{bmatrix} = \begin{bmatrix} \mathbf{0} & \mathbf{I} \\ 0 & A^T \end{bmatrix} \begin{bmatrix} \tilde{y}_{K+1} \\ \vdots \\ \tilde{y}_N \\ \hat{y}_{N+1} \end{bmatrix} = \begin{bmatrix} \mathbf{0} & \mathbf{I} \\ 0 & A^T \end{bmatrix} \begin{bmatrix} \mathbf{0} & \mathbf{I} \\ 0 & A^T \end{bmatrix} \begin{bmatrix} \tilde{y}_K \\ \vdots \\ \tilde{y}_{N-1} \\ \tilde{y}_N \end{bmatrix}, \qquad (4.7)$$

where $\mathbf{0}$ is a vector of zeros of dimension $L-1$, $\mathbf{I}$ is the $(L-1) \times (L-1)$ identity matrix and A^T is the vector as defined in Chapter 1 on page 24. Therefore, using $\widetilde{Z}_j$ to denote the jth column of $\widetilde{\mathbf{X}}$, we can write

$$\widetilde{Z}_{K+h} = \mathbf{W}^h \widetilde{Z}_K, \qquad (4.8)$$

where

$$\mathbf{W}^h = \begin{bmatrix} \mathbf{0} & \mathbf{I} \\ 0 & A^T \end{bmatrix}^h. \qquad (4.9)$$

Let $\omega^{(h)} = (\omega_1^{(h)}, \ldots, \omega_L^{(h)})$ denote the last row of the matrix $\mathbf{W}^h$. Then we can write

$$\hat{y}_{N+h} = \omega^{(h)}\widetilde{Z}_K = \omega_1^{(h)}\tilde{y}_K + \cdots + \omega_L^{(h)}\tilde{y}_N. \quad (4.10)$$

Program (4.2) shows a R code that compute the coefficients for h-step ahead forecasts, $\omega^{(h)}$, by Recurrent SSA method.

Program 4.2 RSSA forecast equation R code

```
RSSA.Eq<-function(L,groups,Y,h){
N<-length(Y)
L<-min(L,(N-L+1))
X<-UniHankel(Y,L)
U<-matrix(svd(X)$u,L,L)
pi<-array(U[L,groups],dim=length(groups))
V2<-sum(pi^2)
m<-length(groups)
Udelta<-array(U[1:(L-1),groups],dim=c((L-1),m))
A<-pi%*%t(Udelta)/(1-V2)
IA<-rbind(diag((L-1)),A)
   W<-cbind(rep(0,L),IA)
   Temp<-diag(L)
   for(i in 1:h){
   Temp<-W%*%Temp
   }
   Temp[L,]
 }
```

4.3.2 Vector SSA Forecast Equation

Considering $Z_{K+c} = [\hat{y}_{K+c}^{(c)}, \ldots, \hat{y}_{N+c}^{(c)}]^T$ for $c = 1, \ldots, h+L-1$ in Eq. (1.15), the vector SSA forecasting approach can be illustrated by Fig. 4.1.

As becomes obvious from Fig. 4.1, the vector SSA forecasting algorithm extracts exactly L values to obtain each forecast, computing each forecast value as the mean of those L values. Similarly to Recurrent SSA it can be concluded from Eq. (1.15) that

$$Z_{K+h} = \mathbf{W}^h Z_K, \quad (4.11)$$

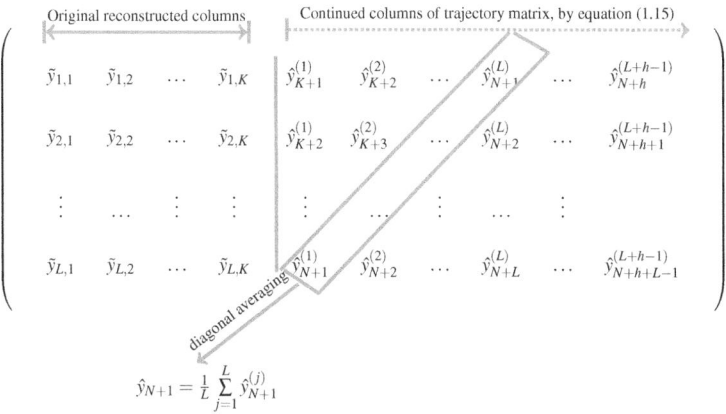

Fig. 4.1 Vector SSA forecasting

where

$$\mathbf{W}^h = \begin{bmatrix} \mathbf{0} & \Pi \\ 0 & A^T \end{bmatrix}^h \quad (4.12)$$

with $\mathbf{0}$ being a vector of dimension $L - 1$ with zero entries, Π and A are defined in Chapter 1. Let $W^h_{(\ell,)}$ denote the ℓth row of $\mathbf{W}^h$, then we can write

$$\hat{y}^{(\ell)}_{N+h} = W^{\ell}_{(L-\ell+h,)} Z_K. \quad (4.13)$$

Therefore, we have:

$$\hat{y}_{N+h} = \frac{1}{L} \sum_{\ell=h}^{h+L-1} W^l_{(L-\ell+h,)} Z_K = \bar{\omega}^{(h)}_1 \tilde{y}_{1,K} + \bar{\omega}^{(h)}_2 \tilde{y}_{2,K} + \cdots + \bar{\omega}^{(h)}_L \tilde{y}_{L,K}, \quad (4.14)$$

where

$$\bar{\omega}^{(h)} = \frac{1}{L} \sum_{\ell=h}^{h+L-1} W^{\ell}_{(L-\ell+h,)}. \quad (4.15)$$

Program (4.3) shows a R code that computes the coefficients for h-step ahead forecasts, $\bar{\omega}^{(h)}$, by vector SSA method.

Program 4.3 VSSA forecast equation R code

```
VSSA.Eq<-function(L,groups,Y,h){
```

```
N<-length(Y)
L<-min(L,(N-L+1))
X<-UniHankel(Y,L)
U<-matrix(svd(X)$u,L,L)
pi<-array(U[L,groups],dim=length(groups))
V2<-sum(pi^2)
m<-length(groups)
Udelta<-array(U[1:(L-1),groups],dim=c((L-1),m))
A<-pi%*%t(Udelta)/(1-V2)
pai<-Udelta%*%t(Udelta)+(1-V2)*t(A)%*%A
Wh<-rep(0,L)
Pv<-rbind(pai,A)
    W<-cbind(rep(0,L),Pv)
    Temp<-diag(L)
    for(ell in 1:(L+h-1)){
    Temp<-W%*%Temp
    b1<-min(L,(L-ell+h))
    b2<-(ell-h+1+abs(ell-h+1))/(ell-h+2+abs(ell-h))
    Wh<-Wh+b2*Temp[b1,]
    }
Wh/L
}
```

Example 4.2: Forecast equation

Consider time series observations in Example 4.1:

```
Y<-c(253.2,313.1,322.7,296.2,370.3,443.8,426.8,453.6,553.4,
+ 588.6, 646.8, 850.0,1106.7,1184.5,1425.1)
```

Then, using $L = 7$ and $r = 2$ and with R code presented in Programs (4.2) and (4.3), we could see the following output for 1 and 5 steps ahead by Recurrent and Vector SSA forecasting methods:

```
> round(RSSA.Eq(L=7,groups=1:2,Y=Y,h=1),3)
[1]  0.000 -0.242 -0.237 -0.120  0.100  0.537  0.970
> round(VSSA.Eq(L=7,groups=1:2,Y=Y,h=1),3)
[1]  0.000 -0.222 -0.216 -0.104  0.106  0.523  0.936
> round(RSSA.Eq(L=7,groups=1:2,Y=Y,h=5),3)
[1]  0.000 -0.670 -1.154 -1.179 -0.557  1.038  3.579
> round(VSSA.Eq(L=7,groups=1:2,Y=Y,h=5),3)
[1]  0.000 -1.434 -1.454 -1.002 -0.146  1.584  3.308
```

Using these results, we can write forecast equations by recurrent method as below:

$$\hat{y}_{16} = 0.970\tilde{y}_{15} + 0.537\tilde{y}_{14} + 0.100\tilde{y}_{13} - 0.120\tilde{y}_{12} - 0.237\tilde{y}_{11} - 0.242\tilde{y}_{10}$$
$$\hat{y}_{20} = 3.579\tilde{y}_{15} + 1.038\tilde{y}_{14} - 0.557\tilde{y}_{13} - 1.179\tilde{y}_{12} - 1.154\tilde{y}_{11} - 0.670\tilde{y}_{10},$$

whereas the results by vector method are as below:

$$\hat{y}_{16} = 0.936\tilde{y}_{7,8} + 0.523\tilde{y}_{6,8} + 0.106\tilde{y}_{5,8} - 0.104\tilde{y}_{4,8} - 0.216\tilde{y}_{3,8} - 0.222\tilde{y}_{2,8}$$
$$\hat{y}_{20} = 3.308\tilde{y}_{7,8} + 1.584\tilde{y}_{6,8} - 0.146\tilde{y}_{5,8} - 1.002\tilde{y}_{4,8} - 1.454\tilde{y}_{3,8} - 1.434\tilde{y}_{2,8}.$$

Rodrigues and Mahmoudvand (2017) presented similar formulas for the multivariate SSA algorithms by using the power of matrices.

4.4 Different Window Length for Forecasting and Reconstruction

The idea of using different window length, say L_{rec} and L_{LRF}, for reconstruction and for forecasting was described in Golyandina (2010) and Mahmoudvand et al. (2013). Golyandina (2010) applies the LRF that was estimated with the window length, L_{LRF}, to the true signal values. She considers the signal with rank r as given in advance, which is not the case in practice. Mahmoudvand et al. (2013) apply the LRF that was estimated with the window length, L_{LRF}, to the original observations. In both approaches, using window length L_{LRF} for forecasting will produce eigenvectors that differ from the eigenvectors provided by L_{rec}. Program (4.4) helps us to compute LRF coefficients.

Program 4.4 LRF coefficients R code

```
LRF<-function(L,groups,Y){
N<-length(Y)
L<-min(L,(N-L+1))
X<-UniHankel(Y,L)
U<-matrix(svd(X)$u,L,L)
pi<-array(U[L,groups],dim=length(groups))
V2<-sum(pi^2)
m<-length(groups)
Udelta<-array(U[1:(L-1),groups],dim=c((L-1),m))
```

```
pi%*%t(Udelta)/(1-V2)
}
```

As an illustration, let us consider series $y_t = s_t + \varepsilon_t$, for $t = 1, \ldots, 20$, where $s_t = t$ and ε_t is white noise with variance one. We compare four scenarios:

(i) $L_{\text{LRF}} = L_{\text{rec}} = 6$ with observations $\{y_t\}$,
(ii) $L_{\text{LRF}} = 5$ and $L_{\text{rec}} \neq 5$ with observations $\{y_t\}$,
(iii) $L_{\text{LRF}} = 5$ with observations $\{s_t\}$ and $L_{\text{rec}} \neq 5$ with observations $\{y_t\}$,

Scenario (i) is the standard approach; scenario (ii) is the approach by Mahmoudvand et al. (2013), and scenario (iii) is the approach proposed by Golyandina (2010). Scenarios (ii) and (iii) are directly comparable as the dimension of LRF is 4 in these scenarios. Using Program (4.4) in the following R code we can compute LRF coefficients for scenarios (i)–(iii).

```
> set.seed(12345)
> st<-1:20
> et<-rnorm(20)
> Yt<-st+et
> LRF(L=6,1:2,Yt)
           [,1]         [,2]        [,3]       [,4]      [,5]
[1,] -0.3828318  -0.09614383  0.1304001  0.5565102  0.7981416
> LRF(L=5,1:2,Yt)
           [,1]         [,2]        [,3]       [,4]
[1,] -0.7095882  0.07259228  0.5259501  1.087332
> LRF(L=5,1:2,st)
      [,1]          [,2]    [,3] [,4]
[1,] -0.5  -1.179612e-15   0.5    1
```

This output confirms the differences between the proposal by Golyandina (2010) and Mahmoudvand et al. (2013).

4.5 OUTLIER IN SSA

The definition of an outlier may be an observation which is unrepresentative, spurious or discordant. It may be regarded as an observation which does not come from the target population. Detecting outliers is important because they have an impact on the selection of the model, the estimation of parameters and, consequently, on forecasts.

In order to detect outliers in SSA, we can fit a model by SSA. Then compute the residuals and compute the following bounds:

$$U = q_{0.75} + 1.5(q_{0.75} - q_{0.25})$$
$$L = q_{0.25} - 1.5(q_{0.75} - q_{0.25}),$$

where $q_{0.25}$ and $q_{0.75}$ are the 25th and 75th percentiles of the residuals, respectively. Outliers are identified as points with residuals larger than U or smaller than L. The following R function can help us to find outliers in a time series:

Program 4.5 Outlier detection R code

```
require(zoo)
outlier<-function(Y,L,groups){
Fitting<-SSA.Rec(Y,L,groups)
res<-Fitting$Residual
approx<-Fitting$Approximation
q25<-quantile(res,.25)
q75<-quantile(res,.75)
low<-q25-1.5*(q75-q25)
up<-q75+1.5*(q75-q25)
Data<-as.zoo(cbind(Y,approx))
out.num<-which(res<low|res>up)
plot.ts(Data[,1],ylim=range(Data[,1:2]),ylab="Time series")
legend("topleft", lty=c(1,2),c("real","fit"),lwd=c(1,2))
lines(Data[,2],lty=2,lwd=2)
if(length(out.num)==0)
print("There is no outlier")
else{
points(Data[out.num,1],pch=19,col="red")
print(out.num)
}
}
```

Example 4.3: Dealing with outliers in SSA

variable death shows dead less than 1 year mortality rate for children-100 born alive in France.

```
> death
      Jan Feb Mar Apr May Jun Jul Aug Sep Oct Nov Dec
```

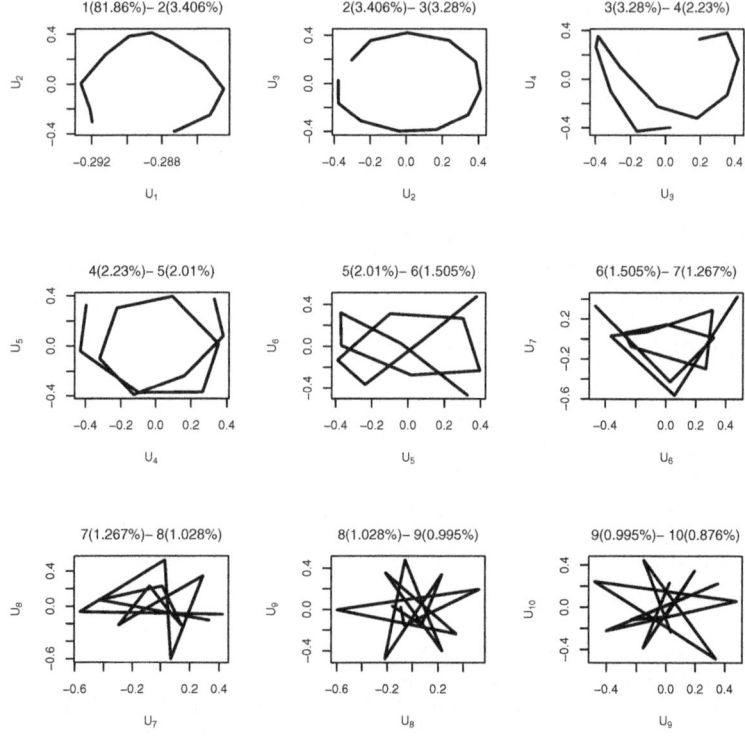

Fig. 4.2 Plot of the first 10 paired eigenvectors for death series

```
1977                                                             13.3
1978 14.5 13.4 11.9 12.0 12.1 12.4 12.2 11.9 12.7 12.9 13.0 13.8
1979 12.6 12.8 11.3 10.9 12.5 12.0 11.8 11.0 11.9 12.5 15.5 14.8
1980 14.3 13.4 13.0 13.1 12.1 12.5 13.3 14.3 13.4 14.0 13.8 13.6
1981 12.1 12.8
```

Using Program (1.8) with $L = 12$, the eigenfunctions for the death series can be obtained by the following command:

`eigen(death,L=12)`

The graphs for the first 10 paired eigenvectors is shown in Fig. 4.2.

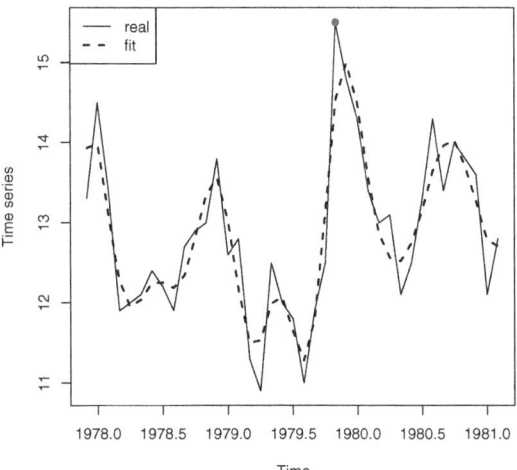

Fig. 4.3 Plot of the death series and fitted curve

This Figure shows that the first five components are enough to construct the series. Applying $L = 12$ and the first five eigentriples with R function outlier we find that the 24th observation is potentially an outlier. This function shows also the plot of time series along with the fitted series and outliers, if there is any, see Fig. 4.3. If there are outliers, they will be identified by red colour in the graph. Here in Fig. 4.3 the 24th point was marked by red colour.

```
> outlier(death,L=12,groups=1:5)
[1] 24
```

Whilst parametric forecasting models have received substantial attention in terms of research into the effects of outliers, the SSA technique has not received such attention. In terms of the effects of outliers, Hassani et al. (2014) gives consideration to both univariate versions of SSA, i.e. the performance of both vector and recurrent forecasting algorithms, the impact on the SVD step, reconstruction stage, and forecasting procedures using simulated and real data. The study also takes into consideration the positioning of the outlier in a time series as this can influence the number of eigenvalues and the SSA forecast.

Appendix A: A Short Introduction to R

A.1 Beginning with R

After R is started, the RGui windows will appear. This window has several parts, of which the R Console window is the main part of RGui where all codes must be typed, directly after the > prompt. Numerical results of calculations with R will be displayed in the R Console; however, graphical output will be depicted in the R Graphics windows. In R, the widely preferred assignment operator is an arrow made from two characters <-, although = can be used instead. For example typing x<-2 will assign number 2 in name x. Similar to other software, there are many predefined functions in R that help users to obtain their results easily. In the next sections, several useful R functions, that will be used here, are presented. In addition, users can get help from the help menu in RGui or type ? before the name of a command or function. For example type ?hist in R console and press enter, and as result a page will appear in which you can find a description of the function hist() with relevant examples.

A.2 Types of Data in R

As in other software, logical, integer, real, complex, string (or character) and raw data can be used in R. These types of data can be defined and saved with different structures. Most applicable structures are:

Vectors

are one-dimensional variables with one or more values of the same type. Vectors can be defined in R by different ways such as:

```
v1<-c(1,2,5)
v1
[1] 1 2 5

v2<-1:5
v2
[1] 1 2 3 4 5

v3<-seq(1,5,0.5)
v3
[1] 1.0 1.5 2.0 2.5 3.0 3.5 4.0 4.5 5.0

v4<-rep(c(1,5),each=2,times=3)
v4
[1] 1 1 5 5 1 1 5 5 1 1 5 5
```

Matrices

are two-dimensional variables with one or more values of the same type. Matrices can be defined using the R function `matrix()` as in the following example:

```
m1<-matrix(c(1,2,5,6,4,9),2,3)
m1
     [,1] [,2] [,3]
[1,]    1    5    4
[2,]    2    6    9
```

Data frames

are similar to matrices, however they have the possibility of including different value types. The R function `data.frame()` produces the data frame, as the following example shows:

```
sex<-c("m","m","f","m","f")
weight<-c(2.5,3.3,2.7,4.1,2.5)
height<-c(45,49,55,37,48)
D<-data.frame(sex,weight,height)
D
```

```
  sex weight height
1   m    2.5     45
2   m    3.3     49
3   f    2.7     55
4   m    4.1     37
5   f    2.5     48
```

Arrays
are one or more dimensional variables with one or more values of the same type. To create an array in R, the function `array()`, can be used as shown in the following example.

```
a<-array(1:24,dim=c(4,3,2))
a
, , 1

     [,1] [,2] [,3]
[1,]    1    5    9
[2,]    2    6   10
[3,]    3    7   11
[4,]    4    8   12

, , 2

     [,1] [,2] [,3]
[1,]   13   17   21
[2,]   14   18   22
[3,]   15   19   23
[4,]   16   20   24
```

Lists
are the most complete forms of data structure, which can include vectors, matrices, data frames, arrays or even other lists. A list in R gives the possibility of having a matrix or array with different lengths for rows or columns. Lists in R can be defined by the function `list()`, as shown in the following example:

```
L1<-c(1,5,3)
L2<-c("A","B")
L<-list(L1,L2)
L
```

```
[[1]]
[1] 1 5 3

[[2]]
[1] "A" "B"
```

As the output of this example shows, the components of list are numbered by `[[1]]`, `[[2]]` and so on. To determine every component, a number must be given component after the name of the list. For instance in the above example we have:

```
L[[1]]
[1] 1 5 3
```

Note that it is possible to assign arbitrary names to the components of lists as demonstrated below:

```
L1<-c(1,5,3)
L2<-c("A","B")
L<-list(Number=L1,Letter=L2)
L
$Number
[1] 1 5 3

$Letter
[1] "A" "B"
```

As indicated, the components numbers have been replaced by names that have been defined. In these cases, the components of lists can be obtained by adding `$name` after the name of the list as shown below:

```
L$Number
[1] 1 5 3
```

A.3 Data Entry in R

Generally speaking, data can be entered into R by using the keyboard or reading from a file. The keyboard method is suitable when there is little data for entering, whereas the data read can be much more extensive. Although it is possible to read data from different extensions of files, such as *.xls (by Excel), *.sav (by SPSS), *.mtb (by Minitab), *.txt (by Notepad), it is preferable to use only *.txt and *.xls, as they are easier to use. For reading data from text and Excel files, it is possible to use the functions `scan()`, `read.txt()` and `read.cls()`.

- `scan()`: to use this function, type it in the command line and then press enter. R will be waiting to receive data from the user. It is then possible to enter data by keyboard and press the enter or space bar key after every filled in data. It is also possible to copy data from excel or notepad files and then paste it in the line where it is possible to enter data.
- `read.table()`: when data is entered in a *.txt file, the following code can be used to read the data of the file:

  ```
  read.table("d:/data.txt")
  ```

  Note that, it is necessary to replace the address of the file instead of d: in the above code.
- `read.csv()`: this function is used for reading data from an Excel file. To do so, the Excel file must first be saved with extension *.csv, then use the following code:

  ```
  read.csv("d:/data.csv")
  ```

A.4 Manipulating and Editing Data in R

One of the limitations of R is related to the editing of commands and data that are typed in the R Console after pressing enter. However, it is possible to access every part of the defined data and edit the value of it. It is also possible to extract one or more parts of data by determining the address of the parts. There are two ways for determining the address of the parts of data that need to be extracted:

- Method 1: by determining the index of the data as in the following examples:

  ```
  x<-c(-2,-1,1,2,4,8)
  x[2]
  [1] -1
  x[2:4]
  ```

```
[1] -1  1  2
x[c(1,3,7)]
[1] -2  1 NA
x[c(1,2,6)]
[1] -2 -1  8
x[-3]
[1] -2 -1  2  4  8
x[-c(1,2)]
[1]  1  2  4  8
```

- Method 2: by defining a rule for choosing the requested data; for the defined vector x, the following codes can be executed:

```
x[x>0]
[1] 1 2 4 8
x[x>0&x<4]
[1] 1 2
x[x^2==4]
[1] -2  2
```

For changing values of one or more parts of the data, it is possible to determine the locations of the data by the above methods and then assign new values, an example of this running code is given below:

```
x[x<0]<-0
```

This running code is to reassign zero to all negative values of the vector x. There are some situations where there are missing values in the data sets and those are indicated by NA.

A.5 Demonstration of R Functions

A.5.1 Computational Functions

A number of computational functions are available in R as in other similar software programs, such as Matlab. Figure A.1 shows the most common computational functions in R. In this figure, grey expressions are the R codes, which must be typed after > prompt in R console whereas the dark expressions next to them indicate the R codes results. For example, typing log(x) gives $Ln(x)$, the natural logarithm of x.

Arithmetic Operators		Logarithm		Big Operators	
x + y	x + y	log(x)	Ln(x)	sum(x[i], i = 1, n)	$\sum_{1}^{n} x_i$
x − y	x − y	log(x, y)	log$_y$(x)	prod(x[i], i = 1, n)	$\prod_{1}^{n} x_i$
x * y	xy	Relations		Statistical Operators	
x/y	x/y	x == y	x = y	mean(x)	$\bar{x}$
Sub/Superscripts		x != y	x ≠ y	var(x)	var(x)
x[i]	x$_i$	x <= y	x ≤ y	min(x)	min x
x^n	x^n	x >= y	x ≥ y	max(x)	max x

Fig. A.1 Some of the computational functions available in R

One of the most prominent capabilities of R, which helps users to do their computations quickly, is acting R functions on the whole element of data structures without using a loop; for example:

```
T<-1:5
log(T)
[1] 0.0000000 0.6931472 1.0986123 1.3862944 1.6094379
```

A.5.2 Graphical Functions

R also has functions that can be used to produce graphs, with plot() the most common R function that can be used to produce graphs.

There is a variety of options designed to assist drawing relevant figures and some of these options are given in the following examples.

```
x<-1:50
plot(sin(2*x*pi/30),type="l",lwd=2,xlab="x",ylab="sin(x)")
plot(cos(2*x*pi/30),lwd=2,xlab="x",ylab="cos(x)")
```

■

These codes will produce the graphs in Fig. A.2.

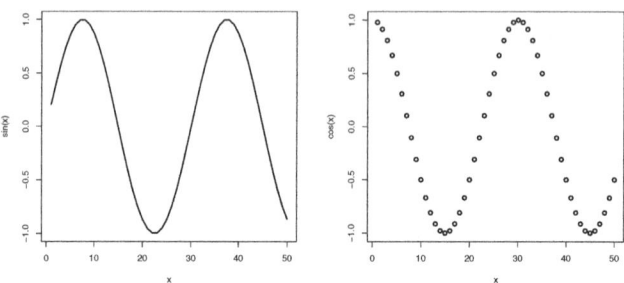

Fig. A.2 Several graphical functions

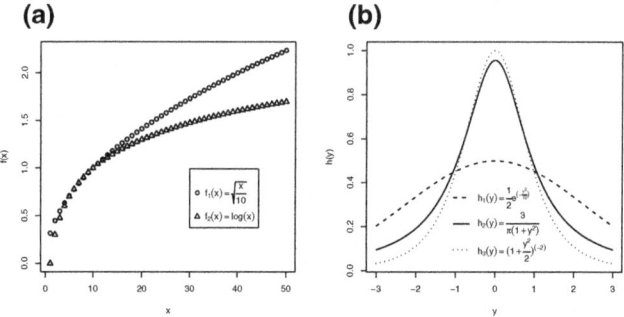

Fig. A.3 Examples of several graphical functions

Using the functions points() and lines() after plot() helps to add several curves in the same plot. Graphs (a) and (b) in Fig. A.3 are produced by these functions with the following codes.

```
x<-1:50
f1<-sqrt(x/10)
f2<-log(x,10)
plot(f1,pch=1,lwd=2,xlab="x",ylab="f(x)",ylim=range(f1,f2))
legend(30,.99,c(expression(f[1](x)==sqrt(frac(x,10))),
expression(f[2](x)==log(x))),pch=c(1,2))
title("(a)")
points(f2,pch=2,lwd=2)

y<-seq(-3,3,length=1000)
h1<-exp(-y^2/10)/2
```

Fig. A.4 Different point characters (pch) for plot function

```
0  1 2 3 4 5 6 7 8 9 10 11 12 13 14 15 16 17 18 19 20 21 22 23 24 25
□  ○ △ + × ◇ ▽ ⊠ * ⊕ ⊙ ⊠ ⊞ ⊠ ⊟ ■ ● ▲ ♦ ● ● ○ □ ◇ △ ▽
```

```
h2<-3/(pi*(1+y^2))
h3<-(1+y^2/2)^(-2)
plot(y,h1,type="l",lty=2,lwd=2,xlab="y",ylab="h(y)",ylim=range(h1,h2,h3))
legend(-1.25,.45,c(expression(h[1](y)==frac(1,2)*e^(-frac(y^2,10))),
expression(h[2](y)==frac(3,pi*(1+y^2))),
expression(h[3](y)==(1+frac(y^2,2))^(-2))),
lty=c(2,1,3),bty="n",lwd=2)
title("(b)")
lines(y,h2,lty=1,lwd=2)
lines(y,h3,lty=3,lwd=2)
```

From the options of plots, pch and lty are useful tools for creating graphs of comparison. For example, Fig. A.4 displays the results of option pch = i for $i = 0, \ldots, 25$ in plot() function. Moreover, line types (lty) can either be specified as an integer (0 = blank, 1 = solid (default), 2 = dashed, 3 = dotted, 4 = dotdash, 5 = longdash, 6 = twodash) or as one of the character strings "blank", "solid", "dashed", "dotted", "dotdash", "longdash", or "twodash", where "blank" uses invisible lines (i.e., does not draw them). A full list of graphical parameters in R are available with ?par.

Multiple Graphs on One Page

The following R functions setup multiple graphs on one page:

- par(mfrow=c(x,y)), and par(mfcol=c(x,y)), where x,y give the number of horizontal and vertical entities, respectively; and figures will be drawn in an *x*-by-*y* array on the device by columns (mfcol), or rows (mfrow), respectively.
- layout(), which allows greater flexibility.

For example, the following R codes generate Fig. A.5:

```
par(mfrow=c(1,3))
hist(rnorm(1000))
boxplot(rnorm(1000))
plot(density(rnorm(1000)))
```

126 APPENDIX A: A SHORT INTRODUCTION TO R

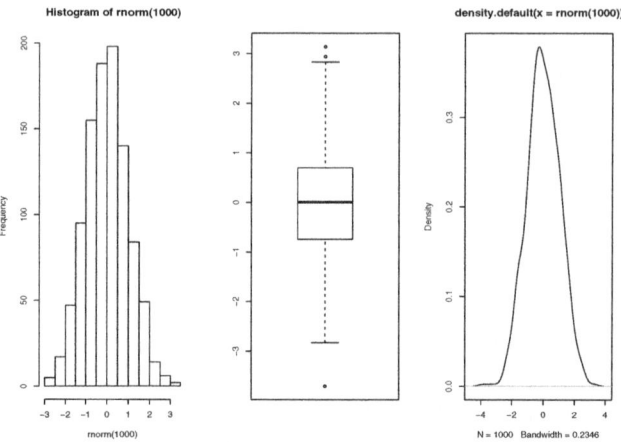

Fig. A.5 Using the mfrow function

Note that the multifigure design, which can be specified by par(), is very rigid. In this function all panels of the plots have the same size; however, in some situations it is better to draw plots with different panel sizes. There is more flexibility with layout(), for instance, it is possible to combine some parts of the page as illustrated in the following R code, which produces Fig. A.6:

```
layout(matrix(c(1,1,2,3),nrow=2,byrow=T))
boxplot(rnorm(1000),horizontal=T)
hist(rnorm(1000))
plot(density(rnorm(1000)))
```

As can be seen in Fig. A.6, the first and second graph's place are combined to produce a better figure than Fig. A.5.

Multiple Graphic Windows

It is useful in some situations to have more than one graphic window at the same time, but note that only one graphics windows can accept graphics commands at any one time, and this is known as the current device. When multiple devices are open, they form a numbered sequence with names giving the kind of device at any position. The R function windows() is one of the main commands used for operating with multiple devices

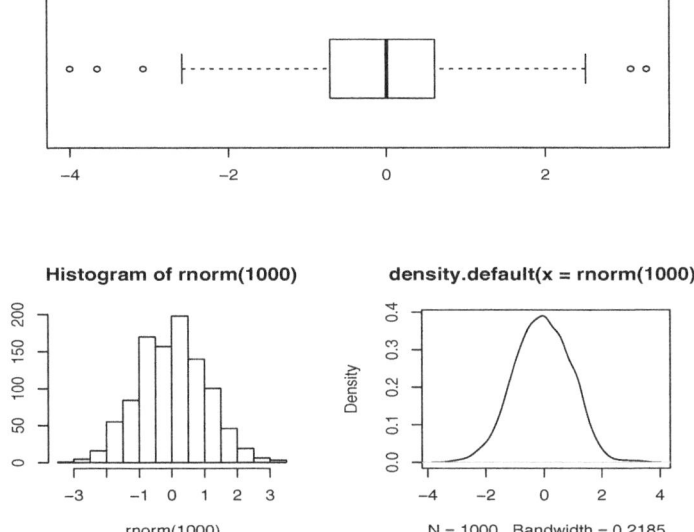

Fig. A.6 Using the layout function

and their meanings. Similar to other R functions, windows() has several different arguments: arguments width, height, which determine the (nominal) width and height of the canvas of the plotting window in inches by default of 7. A full list of arguments can be obtained by ?windows. An example of using windows() for generating a better representation of Fig. A.5 can be constructed by:

```
windows(width=7,height=3)
par(mfrow=c(1,3))
hist(rnorm(1000))
boxplot(rnorm(1000))
plot(density(rnorm(1000))),
```

which produces Fig. A.7.

Saving as Graphs in R

R can generate graphics on almost any type of display such as jpeg, bmp, tiff, gif, pdf and eps. To convert graphical instructions from R into a form mentioned, use the menu File > Save as and then select the

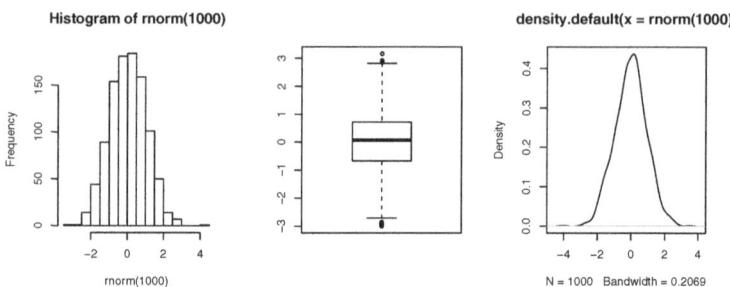

Fig. A.7 Using the mfrow function

suitable format. Another way is using a right click on the graph and choosing a suitable format to export the graph into a file.

A.5.3 Probability Distribution Functions and Their Components in R

Random, density, cumulative and quantile data from known probability distribution functions are the key components, and the way of obtaining them is described here. It is possible to get these values in R by adding the following prefix letters to the name of distribution:

- r: for generating random data.
- d: for computing density function.
- p: for computing cumulative distribution function.
- q: for obtaining quantiles of distribution function.

As an example, the following codes show the output for the normal distribution with mean 0 and standard deviation 1. Note that the statements after # are just explanations for the codes and have not been executed.

```
d1<-dnorm(2,0,1)#computes the density value at point 1.
d1
[1] 0.05399097
p1<-pnorm(1,0,1)#computes the cumulative probability value to the point 1..
p1
[1] 0.8413447
q1<-qnorm(0.975,0,1)#computes the quantile for the level 0.975.
q1
[1] 1.959964
r1<-rnorm(5,0,1)#generates 5 random data.
r1
[1] -0.7007135  0.6981410 -0.3570325  0.8213821  2.2474506
```

A.5.4 Random Samples and Permutations in R

R can also be used to generate arbitrary random data from known distributions. The function `sample()` produces a sample from a vector of numeric data. This function takes a sample of the specified size from the elements of a numeric data vector, either, sampled with or without replacement, as specified in the code. The following example illustrates the use of `sample()` with replacement:

```
x<-c(1,6,2,9,8)
sample(x,10,replace=T)
[1] 1 6 1 1 8 1 1 8 1 2
sample(x,3,replace=F)
[1] 9 8 2
```

A.6 Writing New Functions in R

R also has the option to define new functions or provide some changes as the existing functions. The general form of the new function for R is as follows:

```
funcname<-function(arguments){
body of function
}
```

where `funcname` is an arbitrary name for the function, `arguments` are the input parameters of the function and `body of the function` indicates the computations and output of the function. After defining the function, code `funcname(arguments)`, with given values for arguments, is used to get the results of function. Several examples are provided below.

Example A.1 Assume x is a numeric vector and it is desired to obtain the standardized values of x. This can be achieved with the following function:

```
stand<-function(x){
(x-mean(x))/sd(x)
}
```

After writing this function in the R console, predefined R functions can be used. For example:

```
e<-c(-1.5,-.5,0,2)
stand(e)
[1] -1.0190493 -0.3396831  0.0000000  1.3587324
```

It is possible to get several different outputs from a function. For example, function stand() is changed in such a way that original values, standardized values, mean and variance of data are returned as output.

```
stand<-function(x){
z=(x-mean(x))/sd(x)
data<-data.frame(x,z)
dimnames(data)[[2]]<-c("Original","Standardized")
list(data=data,mean=mean(x),variance=var(x))
}

 stand(e)
$data
  Original Standardized
1     -1.5   -1.0190493
2     -0.5   -0.3396831
3      0.0    0.0000000
4      2.0    1.3587324

$mean
[1] 0

$variance
[1] 2.166667
```

A.6.1 *Loops and Conditions in R*

R functions can be applied to all elements of a vector, matrix or array, without using loops and conditional commands. Although, in some cases, it may be necessary to use a loop or condition to write a new R functions. Note that using loop and condition functions in new functions increases the time of execution and it is better to avoid using loops and conditions wherever possible.

As in other software programs, the function `for()` is the most common loop command in R. The general structure of `for` is as follows:

```
for(counter in setcounter){
commands must be repeated
}
```

Here are several examples:

Example A.2 An identity matrix with dimension 5×5 can be constructed by:

```
I<-matrix(rep(0,25),5,5)
for(i in 1:5)
I[i,i]<-1
I
     [,1] [,2] [,3] [,4] [,5]
[1,]   1    0    0    0    0
[2,]   0    1    0    0    0
[3,]   0    0    1    0    0
[4,]   0    0    0    1    0
[5,]   0    0    0    0    1
```

Example A.3 The centralized moments of a vector of numeric data can be obtained by the following function:

```
moment<-function(x,k){
for(i in 1:k){
M<-mean((x-mean(x))^i)
cat("M",i,"=",M,sep="","\n")}}
```

As an example:

```
x<-1:20
moment(x,10)
M1=0
M2=33.25
M3=0
M4=1983.362
M5=0
```

Table A.1 A discrete distribution

Value	x_1	x_2	x_k
Prob	p_1	p_2	p_k

```
M6=140371.7
M7=0
M8=10781840
M9=0
M10=868289604
```

The R function if() is the usual conditional command like as in other software programs.

```
if(cond1) expr1
else expr2
```

Example A.4 Here we write a function that produces the power of a square matrix. The following code is an example for obtaining the power of a matrix.

```
M.power<-function(M,p){
size<-nrow(M)
M1<-diag(rep(1,size))
if(p==1)
return(M)
else{
for(i in 1:p)
M1<-M1%*%M
return(M1)
}}
```

Example A.5 In this example, the new function produces a sample with size *n* from the following discrete distribution.

To solve this problem, first recall that generating data from such distributions is possible by the following steps:

- Generate samples $u_1, \ldots, u_n$ from the uniform $(0, 1)$ distribution,
- Let $p_1 + \ldots + p_{i-1} < u_j \leq p_1 + \ldots + p_i$ then x_i is a random sample from the given distribution.
- Do the previous step for $j = 1, \ldots, n$ and determine $x_1, \ldots, x_n$

Applying this algorithm, the following function generates random samples from the specified discrete distribution:

```
rdisc<-function(n,p,x){
k<-length(x)
ran<-array(dim=n)
u<-runif(n)
P<-cumsum(p)
for(i in 1:n){
if(u[i]<P[1])
ran[i]<-x[1]
else{
for(j in 1:(k-1)){
if(u[i]<P[(j+1)]& u[i]>P[j])
ran[i]<-x[(j+1)]
}}}
ran
}
```

To use `rdisc()` vectors x, p and sample size n must be determined. For instance it is possible to use the following code:

```
p<-c(.1,.3,.4,.2)
x<-c(-1,0,1,2)
rdisc(10,p,x)
[1]  1  2  0  1 -1  1  1  1  0 -1
```

Example A.6 A colour scaling plot is useful in certain situations to show the rate of association between two variables. The following code is an example showing how a scaling plot can be created:

```
Plt.Img <- function(x){
min <- min(x)
max <- max(x)
yLabels <- rownames(x)
xLabels <- colnames(x)
```

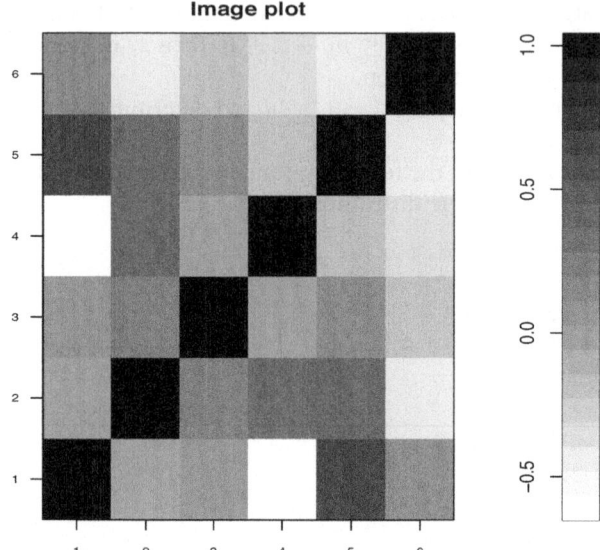

Fig. A.8 Examples of several graphical functions

```
if( is.null(xLabels) ){
xLabels <- c(1:ncol(x))
}
if( is.null(yLabels) ){
yLabels <- c(1:nrow(x))
}
layout(matrix(data=c(1,2), nrow=1, ncol=2),
widths=c(4,1), heights=c(1,1))
ColorRamp <- gray( seq(1,0,length=20))
ColorLevels <- seq(min, max, length=length(ColorRamp))
par(mar = c(3,5,2.5,2))
image(1:length(xLabels), 1:length(yLabels),
t(x), col=ColorRamp, xlab="",
ylab="", axes=FALSE, zlim=c(min,max))
title(main=c("Image plot"))
axis(BELOW<-1, at=1:length(xLabels),
labels=xLabels, cex.axis=0.7)
axis(LEFT <-2, at=1:length(yLabels),
labels=yLabels, las= HORIZONTAL<-1,
cex.axis=0.7)
```

```
box()
par(mar = c(3,2.5,2.5,2))
image(1, ColorLevels,
matrix(data=ColorLevels, ncol=length(ColorLevels),nrow=1),
col=ColorRamp,
xlab="",ylab="",
xaxt="n")
layout(1)
}
```

Now considering matrix *m* as below, the results of function Plt.Img() is similar to w-correlation plot used in SSA:

```
y<-matrix(rbinom(60,10,.5),12,5)
m<- cor(y)
Plt.Img(m)
```

Appendix B: Theoretical Explanations

In this appendix, we provide details of a number of theorems referred in Chapters 1 and 2. We also use the opportunity to elaborate some concepts used in the previous chapters.

Singular Value Decomposition (SVD)

The singular value decomposition (SVD) is a factorization of a real or complex matrix. It is the generalization of the eigendecomposition of a positive semidefinite normal matrix (e.g., a symmetric matrix with positive eigenvalues) to any $L \times K$ matrix $\mathbf{X}$ via an extension of polar decomposition.

The SVD of $L \times K$ matrix $\mathbf{X}$ is a factorization of the form $\mathbf{U}\mathbf{\Sigma}\mathbf{V}^T$, where $\mathbf{U}$ is an $L \times L$ real or complex unitary matrix, $\mathbf{\Sigma}$ is a $L \times K$ rectangular diagonal matrix with non-negative real numbers on the diagonal, and $\mathbf{V}$ is a $K \times K$ real or complex unitary matrix. The diagonal entries λ_i of $\mathbf{\Sigma}$ are known as the singular values of $\mathbf{X}$. The columns of $\mathbf{U}$ and the columns of $\mathbf{V}$ are called the left-singular vectors and right-singular vectors of $\mathbf{X}$, respectively.

Note that the left-singular values of $\mathbf{X}$ are a set of orthonormal eigenvectors of $\mathbf{X}\mathbf{X}^T$, the right-singular vectors of $\mathbf{X}$ are a set of orthonormal eigenvectors of $\mathbf{X}^T\mathbf{X}$, and the nonzero singular values of $\mathbf{X}$ (which are on the diagonal entries of $\mathbf{\Sigma}$) are the square roots of the nonzero eigenvalues of both $\mathbf{X}\mathbf{X}^T$ and $\mathbf{X}^T\mathbf{X}$.

138   APPENDIX B: THEORETICAL EXPLANATIONS

Theorem B.1 *Let d denotes the rank of the trajectory matrix $\mathbf{X}$ in SSA. The maximum rank of the trajectory matrix $\mathbf{X}$ is achieved at $L = L_{\max}$.*

Proof The maximum value of d is obtained as follows:

$$\max_{L \in \{2,\ldots,N-1\}} d = \max_{L \in \{2,\ldots,N-1\}} \min\{L, N - L + 1\}$$
$$= \max_{L \in \{2,\ldots,N-1\}} \frac{N + 1 - 2L - (N+1)}{2},$$

which shows that the maximum rank of $\mathbf{X}$ is attained when $2L - (N+1)$ is minimized. Thus, $L = L_{\max}$.   ∎

Theorem B.2 *Let d denote the rank of the trajectory matrix $\mathbf{X}$ in multivariate SSA (either VMSSA or HMSSA). The maximum rank of the trajectory matrix $\mathbf{X}$ is achieved at $L = \left\lceil \frac{1}{M+1}(N+1) \right\rceil$ and $L = \left\lceil \frac{M}{M+1}(N+1) \right\rceil$ for VMSSA and HMSSA, respectively, where $\lceil u \rceil$ denotes the closest integer number to u.*

Proof Let us, for example, consider the VMSSA version. The maximum value of d is obtained as follows:

$$\max d = \max_{L \in \{2,\ldots,N-1\}} \min\{LM, N - L + 1\}$$
$$= \max_{L \in \{2,\ldots,N-1\}} \frac{L(M-1) + N + 1 - |L(M+1) - (N+1)|}{2} \quad (B.1)$$

which shows that the maximum rank of $\mathbf{X}$ is attained when $|L(M+1) - (N+1)|$ is minimized. Thus, $L = \left\lceil \frac{N+1}{M+1} \right\rceil$. Similarly, it is straightforward to prove that $L = \left\lceil \frac{M}{M+1}(N+1) \right\rceil$ provides the optimal L for VMSSA. Moreover, these values of L are the first $(M+1)$-quantile and the Mth $(M+1)$-quantile of the integer set $\{1, \ldots, N\}$. The optimal values of L in the reconstruction stage of MSSA, having M series with length N, are as follows:

$$L = \begin{cases} \left\lceil \frac{1}{M+1} \right\rceil (N+1) & \text{VMSSA;} \\ \left\lceil \frac{M}{M+1} \right\rceil (N+1) & \text{HMSSA.} \end{cases} \quad (B.2)$$

∎

Theorem B.3 *Consider the Hankel matrix* $\mathbf{X}$ *as defined in Eq. (1.2), then,*

$$T_\mathbf{X}^{L,N} = \sum_{j=1}^{N} w_j^{L,N} y_j^2,$$

where $w_j^{L,N} = \min\{\min\{L, K\}, j, L + K - j\} = w_j^{K,N}$.

Proof Applying the definition of $\mathbf{X}$, we have:

$$T_\mathbf{X}^{L,N} = \sum_{i=1}^{L} \sum_{j=i}^{N-L+i} y_j^2. \tag{B.3}$$

Changing the order of the summations in Eq. (B.3), we observe

$$T_\mathbf{X}^{L,N} = \sum_{j=1}^{N} C_{j,L,N} y_j^2,$$

where $C_{j,L,N} = \min\{j, L\} - \max\{1, j - N + L\} + 1$. We only need therefore to show that $C_{j,L,N} = w_j^{L,N}$, for all j and L. We consider two cases: $L \leq K$ and $L > K$. For the first case, we have

$$C_{j,L,N} = \begin{cases} j, & 1 \leq j \leq L, \\ L, & L + 1 \leq j \leq K, \\ N - j + 1, & K + 1 \leq j \leq N, \end{cases}$$

which is exactly equal to $w_j^{L,N}$. Similarly for the second case, we have

$$C_{j,L,N} = \begin{cases} j, & 1 \leq j \leq K, \\ K, & K + 1 \leq j \leq L, \\ N - j + 1, & L + 1 \leq j \leq N, \end{cases}$$

and again is equal to $w_j^{L,N}$, for $L > K$. ∎

Theorem B.4 *Let* $T_\mathbf{X}^{L,N}$ *be defined as before, then,* $T_\mathbf{X}^{L,N}$ *is an increasing function of* L *on* $\{2, \ldots, [(N + 1)/2]\}$ *and a decreasing function on* $\{[(N + 1)/2] + 1, \ldots, N - 1\}$, *and*

$$\max T_\mathbf{X}^{L,N} = T_\mathbf{X}^{\left[\frac{N+1}{2}\right],N}.$$

Proof First, we show that $w_j^{L,N}$ is an increasing function of L on $\{2, \ldots, [(N+1)/2]\}$. Let L_1 and L_2 be two arbitrary values where $L_1 < L_2 \leq [(N+1)/2]$. From the definition of $w_j^{L,N}$, we have

$$w_j^{L_2,N} - w_j^{L_1,N} = \begin{cases} 0, & 1 \leq j \leq L_1, \\ j - L_1, & L_1 + 1 \leq j \leq L_2, \\ L_2 - L_1, & L_2 + 1 \leq j \leq N - L_2 + 1, \\ N - j + 1 - L_1, & N - L_2 + 2 \leq j \leq N - L_1 + 1, \\ 0, & N - L_1 + 2 \leq j \leq N. \end{cases}$$

Therefore, $w_j^{L_2,N} - w_j^{L_1,N} \geq 0$, for all j, and the inequality is strict for some j. Thus,

$$T_{\mathbf{X}}^{L_2,N} - T_{\mathbf{X}}^{L_1,N} = \sum_{j=1}^{N} \left(w_j^{L_2,N} - w_j^{L_1,N} \right) h_j^2 > 0. \quad (B.4)$$

This confirms that w_j^L is an increasing function of L on $\{2, \ldots, [(N+1)/2]\}$. A similar approach for the set $\{[(N+1)/2]+1, \ldots, N-1\}$ implies that $w_j^{L,N}$ is a decreasing function of L on this interval. Note also that $T_{\mathbf{X}}^{L_2,N} - T_{\mathbf{X}}^{L_1,N}$ in Eq. (B.4) increases as we increase the value of L_2 proving that $T_{\mathbf{X}}^{L,N}$ is an increasing function on $\{2, \ldots, [(N+1)/2]\}$. The maximum value of $T_{\mathbf{X}}^{L_2,N}$ is therefore attained at the maximum value of L, which is $[(N+1)/2]$. ∎

Corollary B.1 *Let $L_{\max}$ denote the value of L such that $T_{\mathbf{X}}^{L,N} \leq T_{\mathbf{X}}^{L_{\max},N}$, for all L, and the inequality to be strict for some values of L then,*

$$L_{\max} = \begin{cases} \frac{N+1}{2}, & \text{if } N \text{ is odd}, \\ \frac{N}{2}, \frac{N}{2} + 1, & \text{if } N \text{ is even}. \end{cases}$$

Corollary B.1 shows that $L = \text{median}\{1, \ldots, N\}$ maximizes the sum of squares of the Hankel matrix singular values with fixed values of N. Applying Corollary B.1 and the definition of $w_j^{L,N}$, we can show that

$$w_j^{L_{\max},N} = \frac{N+1}{2} - \left| \frac{N+1}{2} - j \right|. \quad (B.5)$$

Equation (B.5) shows that $y_{[(N+1)/2]}$ has maximum weight at $T_{\mathbf{X}}^{L,N}$.

Theorem B.5 $T_X^{L,N}/K$ *is an increasing function of L on $\{2, ..., N-1\}$.*

Proof L_1 and L_2 are two arbitrary values where $L_1 < L_2$. It is necessary to show that $K_1 T_X^{L_2,N} - K_2 T_X^{L_1,N} > 0$, where $K_j = N - L_j + 1$ ($j = 1, 2$). To this end, the following four cases are considered:

- Case 1: $L_2' \leq L_{\max}$. In this case, the results are easily obtained from Theorem B.4.
- Case 2: $L_1 \leq L_{\max} < L_2 < N - L_1 + 1$. Using the equality $T_X^{L_2,N} = T_X^{N-L_2+1,N}$ can be obtained. Now, applying the inequality $N - L_2 + 1 > L_1$ for this case, the results are again obtained from Theorem B.4.
- Case 3: $L_1 \leq L_{\max} \leq N - L_1 + 1 < L_2$. In this case:

$$K_1 w_j^{L_2,N} - K_2 w_j^{L_1,N} = \begin{cases} j(L_2 - L_1), & 1 \leq j \leq K_2, \\ K_2(K_1 - j), & K_2 + 1 \leq j \leq L_1, \\ K_2(K_1 - L_1), & L_1 + 1 \leq j \leq K_1, \\ K_2(j - L_1), & K_1 + 1 \leq j \leq L_2, \\ (N - j + 1)(L_2 - L_1), & L_2 + 1 \leq j \leq N, \end{cases}$$

Therefore, $K_1 w_j^{L_2,N} - K_2 w_j^{L_1,N} > 0$ for all j, which gives the required result.

- Case 4: $L_{\max} \leq L_1 < L_2$. Similar to the case 3 there is:

$$K_1 w_j^{L_2,N} - K_2 w_j^{L_1,N} = \begin{cases} j(L_2 - L_1), & 1 \leq j \leq K_2, \\ K_2(K_1 - j), & K_2 + 1 \leq j \leq K_1, \\ 0, & K_1 + 1 \leq j \leq L_1, \\ K_2(j - L_1), & L_1 + 1 \leq j \leq L_2, \\ (N - j + 1)(L_2 - L_1), & L_2 + 1 \leq j \leq N, \end{cases}$$

Therefore, $K_1 w_j^{L_2,N} - K_2 w_j^{L_1,N} \geq 0$ for all j and inequality is strict for some j. ∎

Theorem B.6 *The matrix Π is the matrix of the linear operator that performs the orthogonal projection $\mathbb{R}^{L_{sum}-M} \mapsto \mathcal{L}_r^\nabla$, where $\mathcal{L}_r^\nabla = Span\{U_1^\nabla, \ldots, U_r^\nabla\}$.*

Proof Let $\mathbf{U}$ be a matrix with columns $U_1, \ldots, U_r$. By definition of U_j we can write:
$$\mathbf{I}_{r \times r} = \mathbf{U}^T \mathbf{U} = \mathbf{U}^{\nabla T} \mathbf{U}^\nabla + \mathbf{W}^T \mathbf{W}. \tag{B.6}$$
Thus, $\mathbf{U}^{\nabla T} \mathbf{U}^\nabla = \mathbf{I}_{r \times r} - \mathbf{W}^T \mathbf{W}$. Using Eq. (B.6), we can write:

$$\left(\mathbf{U}^{\nabla T} \mathbf{U}^\nabla\right) \left(\mathbf{I}_{r \times r} + \mathbf{W}^T \left(\mathbf{I}_{M \times M} - \mathbf{W}\mathbf{W}^T\right)^{-1} \mathbf{W}\right)$$
$$= \left(\mathbf{I}_{r \times r} - \mathbf{W}^T \mathbf{W}\right) \left(\mathbf{I}_{r \times r} + \mathbf{W}^T \left(\mathbf{I}_{M \times M} - \mathbf{W}\mathbf{W}^T\right)^{-1} \mathbf{W}\right)$$
$$= \mathbf{I}_{r \times r} - \mathbf{W}^T \mathbf{W} + \mathbf{W}^T \left(\mathbf{I}_{M \times M} - \mathbf{W}\mathbf{W}^T\right) \left(\mathbf{I}_{M \times M} - \mathbf{W}\mathbf{W}^T\right)^{-1} \mathbf{W} = \mathbf{I}_{r \times r}$$

Therefore $\left(\mathbf{U}^{\nabla T} \mathbf{U}^\nabla\right)^{-1} = \left(\mathbf{I}_{r \times r} + \mathbf{W}^T \left(\mathbf{I}_{M \times M} - \mathbf{W}\mathbf{W}^T\right)^{-1} \mathbf{W}\right)$ and the perpendicular projection onto the column space of $\mathbf{U}^\nabla$ can be obtained by the following equation:

$$\mathbf{U}^\nabla \left(\mathbf{U}^{\nabla T} \mathbf{U}^\nabla\right)^{-1} \mathbf{U}^{\nabla T} = \mathbf{U}^\nabla \mathbf{U}^{\nabla T} + \mathbf{U}^\nabla \mathbf{W}^T \left(\mathbf{I}_{M \times M} - \mathbf{W}\mathbf{W}^T\right)^{-1} \mathbf{W} \mathbf{U}^{\nabla T}, \tag{B.7}$$

which completes the proof by using the definition of $\mathcal{R}$. ∎

Strong separability

Fix the window length L, consider a certain SVD of the trajectory matrix $\mathbf{X}$ of the initial series Y_N of length N, and assume that this series is a sum of two series $Y_N^{(1)}$ and $Y_N^{(2)}$, that is, $Y_N = Y_N^{(1)} + Y_N^{(2)}$. Moreover, let $\mathbf{X}^{(1)}$ and $\mathbf{X}^{(2)}$ be the corresponding trajectory matrices of the series $Y_N^{(1)}$ and $Y_N^{(2)}$, respectively.

Separability of the series $Y_N^{(1)}$ and $Y_N^{(2)}$ means that the matrix terms of the SVD can be split from the trajectory matrix $\mathbf{X}$ into two different groups, so that the sums of terms within the groups give the trajectory matrices $\mathbf{X}^{(1)}$ and $\mathbf{X}^{(2)}$ of the series $Y_N^{(1)}$ and $Y_N^{(2)}$, respectively. In other words assume $\mathbf{X} = \mathbf{X}_{SVD}^{(1)} + \mathbf{X}_{SVD}^{(2)}$ by using SVD and grouping. Then strong separability occurs if $\mathbf{X}_{SVD}^{(1)} = \mathbf{X}^{(1)}$ and $\mathbf{X}_{SVD}^{(2)} = \mathbf{X}^{(2)}$.

Although, strong separability only rarely occurs in practice; it happens theoretically if the columns (rows) of $\mathbf{X}^{(1)}$ and $\mathbf{X}^{(2)}$ are orthogonal and the singular values are isolated. Zokaei and Mahmoudvand (2011) have considered the coefficient of variation for the differentiation of singular values and showed that the maximum separation for a wide class of series will be obtained when $L = L_{\max}$.

References

Alonso, F., Castillo, J., & Pintado, P. (2005). Application of singular spectrum analysis to the smoothing of raw kinematic signals. *Journal of Biomechanics, 38*(2), 1085–1092.

Alvarez-Meza, A., Acosta-Medina, C., & Castellanos-Domnguez, G. (2013). Automatic singular spectrum analysis for time-series decomposition. In *ESANN 2013 Proceedings, European Symposium on Artificial Neural Networks, Computational Intelligence and Machine Learning* (pp. 131–136). Available from http://www.i6doc.com/en/livre/?GCOI=28001100131010.

Broomhead, D. S., & King, G. P. (1986a). Extracting qualitative dynamics from experimental data. *Physica D, 20*, 217–236.

Broomhead, D. S., & King, G. P. (1986b). *On the qualitative analysis of experimental dynamical systems*. Bristol: Adam Hilger.

Broomhead, D. S., King, G. P., & Pike, E. R. (1987). *Singular spectrum analysis with application to dynamical systems*. Bristol: IOP Publication.

Elsner, J. B., & Tsonis, A. A. (1996). *Singular spectrum analysis: A new tool in time series analysis*. New York: Plenum Press.

Ghodsi, M., Hassani, H., Sanei, S., & Hichs, Y. (2009). The use of noise information for detection of temporomandibular disorder. *Biomedical Signal Processing and Control, 4*(2), 79–85.

Golyandina, N. (2010). On the choice of parameters in singular spectrum analysis: And related subspace-based methods. *Statistics and Inference, 3*(3), 257–279.

Golyandina, N., Nekrutkin, V., & Zhiglovsky, A. (2001). *Analysis of time series structure: SSA and related techniques*. London: Chapman & Hall/CRC.

Golyandina, N., & Osipov, E. (2007). The caterpillar-SSA method for analysis of time series with missing values. *Journal of Statistical Planning and Interface, 137*(8), 2642–2653.

Golyandina, N., & Zhigljavsky, A. (2013). *Singular spectrum analysis for time series*. Springer brief in statistics. Berlin: Springer.

Hamilton, J. D. (2017). Why you should never use the Hodrick–Prescott filter. *Review of Economics and Statistics*.

Harris, T., & Yan, H. (2010). Filtering and frequency interpretations of singular spectrum analysis. *Physica D, 239,* 1958–1967.

Hassani, H. (2010). Singular spectrum analysis based on the minimum variance estimator. *Nonlinear Analysis: Real World Applications, 11*(3), 2065–2077.

Hassani, H., & Mahmoudvand, R. (2013). Multivariate singular spectrum analysis: A general view and new vector forecasting algorithm. *International Journal of Energy and Statistics, 1*(1), 55–83.

Hassani, H., Mahmoudvand, R., Nabe Omer, H., & Silvia, E. S. (2014). A preliminary investigation into the effect of outlier(s) on singular spectrum analysis. *Fluctuation and Noise Letters, 13*(4), https://doi.org/10.1142/S0219477514500291.

Hassani, H., Mahmoudvand, R., & Yarmohammadi, M. (2010). Filtering and denoising in the linear regression model. *Fluctuation and Noise Letters, 9*(4), 343–358.

Hassani, H., Mahmoudvand, R., & Zokaei, M. (2011a). Separability and window length in the singular spectrum analysis. *Compets Rendes Mathematiques, 349,* 987–990.

Hassani, H., Mahmoudvand, R., Zokaei, M., & Ghodsi, M. (2012). On the separability between signal and noise in singular spectrum analysis. *Fluctuation and Noise Letters, 11*(2), 1–11.

Hassani, H., & Thomakos, D. (2010). A review on singular spectrum analysis for economic and financial time series. *Statistics and Its Interface, 3*(3), 377–397.

Hassani, H., Zhigljavsky, A., & Zhengyuan, X. (2011b). Singular spectrum analysis based on the perturbation theory. *Nonlinear Analysis: Real World Applications, 2*(5), 2752–2766.

Khan, M. A. R., & Poskitt, D. S. (2010). *Description length based signal detection in singular spectrum analysis* (Working Paper).

Khan, M. A. R., & Poskitt, D. S. (2013a). Moment tests for window length selection in singular spectrum analysis of short and long memory processes. *Journal of Time Series Analysis, 34*(2), 141–155.

Khan, M. A. R., & Poskitt, D. S. (2013b). A note on window length selection in singular spectrum analysis. *Australian & New Zealand Journal of Statistics, 55*(2), 87–108.

Kondrashov, D., & Ghil, M. (2006). Spatio temporal filling of missing points in geophysical data sets. *Nonlin Processes Geophys, 13,* 151–159.

Kume, K., & Nose-Togawa, N. (2014). Multidimensional extension of singular spectrum analysis based on filtering interpretation. *Advances in Adaptive Data Analysis, 6,* 1–17.

Mahmoudvand, R., Najari, N., & Zokaei, M. (2013). On the optimal parameters for reconstruction and forecasting in the singular spectrum analysis. *Communication in Statistics: Simulations and Computations, 42,* 860–870.

Mahmoudvand, R., & Rodrigues, P. (2016a). Correlation analysis in contaminated data by singular spectrum analysis. *Quality and Reliability Engineering International*, 32, 2127–2137.

Mahmoudvand, R., & Rodrigues, P. (2016b). Missing value imputation in time series using singular spectrum analysis. *International Journal of Energy and Statistics*, 4, 1650005 (6 pages).

Mahmoudvand, R., & Zokaei, M. (2012). On the singular values of the Hankel matrix with application in singular spectrum analysis. *Chilean Journal of Statistics*, 3(1), 43–56.

Mohammad, Y., & Nishida, T. (2011). On comparing SSA-based change point discovery algorithms. *IEEE SII* (pp. 938–945).

Moskvina, V., & Zhigljavsky, A. (2003). An algorithm based on singular spectrum analysis for change-point detection. *Communication in Statistics: Simulation and Computation*, 32(2), 319–352.

Patterson, K., Hassani, H., Heravi, S., & Zhigljavsky, A. (2011). Forecasting the final vintage of the index of industrial production. *Journal of Applied Statistics*, 38(10), 2183–2211.

Rendall, R., & Reis, M. (2014). A comparison study of single-scale and multi-scale approaches for data-driven and model-based online denoising. *Quality and Reliability Engineering International*, 30(7), 935–950.

Rodrigues, P., & de Carvalho, M. (2013). Spectral modeling of time series with missing data. *Applied Mathematical Modelling*, 37, 4676–4684.

Rodrigues, P., & Mahmoudvand, R. (2017). The benefits of multivariate singular spectrum analysis over the univariate version. *Journal of the Franklin Institute*, 355, 544–564.

Sanei, S., & Hassani, H. (2015). *Singular spectrum analysis of biomedical signals*. Boca Raton: CRC Press.

Schoellhamer, D. (2001). Singular spectrum analysis for time series with missing data. *Geophysical Research Letters*, 28(16), 3187–3190.

Shen, Y., Peng, F., & Li, B. (2015). Improved singular spectrum analysis for time series with missing data. *Nonlinear Processes in Geophysics*, 22, 371–376.

INDEX

A
Accuracy measure, 28
Array, 119
Automated MSSA, 79, 80, 82
Automated SSA, 33
Autoregressive, 27

B
Bivariate time series, 52
Block Hankel matrix, 55, 64
Bootstrap, 38, 39

C
Change point, 87
Coefficient of variation, 143
Confidence level, 39
Correlation, 97
Covariance matrix, 15, 17, 60, 65

D
Data.frame, 118
Decomposition, 3, 5, 50, 59
Denoising, 96

Diagonal averaging, 8, 10, 34, 60
Dissimilarity, 65

E
Eigenfunction, 24
Eigentriple, 22, 29, 70
Eigenvalue, 5, 7, 12, 64
Eigenvector, 5, 26, 52
Embedding, 3, 7
Ergodic, 39
Error, 28, 39
Estimation, 15, 39, 93

F
Filtering, 3, 50, 96
Forecasting, 2, 27, 50, 68, 92

G
Gap filling, 92
Grouping, 7, 9, 15, 22, 55

H
Hankel matrix, 4, 5, 8, 139

Hankelization, 9, 60
Horizontal MSSA, 50

I
Imputation, 92

L
Lagged vectors, 17, 65
Linear Algebra, 4
Linear projection, 77
Linear recurrent formula, 27, 69, 73
Linearity, 1
List, 119
Loss function, 33, 80

M
Median, 16
Minimum variance, 2
Multivariate SSA, 2, 49, 50

N
Noise, 3, 5, 21, 50
Nonlinearity, 1
Non-parametric, 1
Normality, 1

O
Orthogonality, 64
Orthogonal projection, 142
Oscillation, 3, 15
Outlier, 20, 30, 96

P
Periodogram, 8
Perturbation, 2
Prediction interval, 38
Principal component, 53

Q
Quantile, 39

R
Random sample, 129
Random variable, 3, 21
Rank, 16, 55, 60, 65, 138
Reconstruction, 3, 7, 15, 50, 92
Recurrent SSA, 28, 29
Regression, 96
Residual, 10, 38, 39, 57, 62
R function, 11, 122, 129
Root Mean Squared Error, 28, 80

S
Seasonality, 2, 12
Separability, 15, 17, 21, 142
Signal, 2, 5, 42, 45
Similarity, 64, 65
Simulation, 20, 28, 39, 65, 98
Singular value, 5, 15, 23, 41, 44, 69, 80
Singular Value Decomposition, 5, 9
Smoothing, 3
Spectrum, 5
Stationarity, 1, 39
Stochastic process, 3
Strong separability, 17, 142

T
Time series, 1
Trace, 17, 65
Trajectory matrix, 3, 22, 30, 50, 64, 93, 138
Transformation, 2, 8, 55
Trend, 3

U
Uniform distribution, 5
Univariate SSA, 56, 60, 68

V

Variance, 2, 23
Vector SSA, 28, 30
Vertical MSSA, 50, 59

W

w-correlation, 15, 17, 41, 45
Weak separability, 18
Window length, 3, 15, 22, 64, 142
w-orthogonal, 18

The manufacturer's authorised representative in the EU is Springer Nature Customer Service Centre GmbH, Europaplatz 3, 69115 Heidelberg, Germany. If you have any concerns regarding our products, please contact ProductSafety@springernature.com

Printed and bound by CPI Group (UK) Ltd, Croydon, CR0 4YY

23/03/2026

02076355-0009